"*Starlings* is a smart, entertaining parable about human foolishness, avian ingenuity, and the unintended consequences of ecological meddling. With wit and verve, Mike Stark tells the epic story of the plucky starling—a bird that enchanted Mozart, exasperated farmers, and ultimately conquered America."

—BEN GOLDFARB, author of *Crossings* and *Eager*

"Americans have been bewitched, befuddled, and enraged by European starlings for more than a century, and the country's least-loved nonnative bird couldn't ask for a better chronicler than Mike Stark. Balanced, whimsical, and deeply researched, *Starlings* tells the story of how they became the bird we love to hate, and in doing so illuminates our own contradictory human nature."

—MELISSA L. SEVIGNY, author of *Brave the Wild River*

"In *Starlings* Mike Stark peels back 150 years of myth and misunderstanding to reveal a fascinating story about human folly, animal smarts, and the value of life on Earth. You'll never look at a starling the same way again."

—MICHELLE NIJHUIS, author of *Beloved Beasts: Fighting for Life in an Age of Extinction*

publication supported by a grant from
The Community Foundation for Greater New Haven
As part of the Urban Haven Project

Starlings
The Curious Odyssey of a Most Hated Bird

MIKE STARK

University of Nebraska Press
LINCOLN

© 2025 by the Board of Regents of the University of Nebraska

All rights reserved

The University of Nebraska Press is part of a land-grant institution with campuses and programs on the past, present, and future homelands of the Pawnee, Ponca, Otoe-Missouria, Omaha, Dakota, Lakota, Kaw, Cheyenne, and Arapaho Peoples, as well as those of the relocated Ho-Chunk, Sac and Fox, and Iowa Peoples.

Library of Congress Cataloging-in-Publication Data
Names: Stark, Mike (Former journalist), author.
Title: Starlings: the curious odyssey of a most hated bird / Mike Stark.
Description: Lincoln: University of Nebraska Press, [2025]. | Includes bibliographical references.
Identifiers: LCCN 2024022118
ISBN 9781496242020 (paperback)
ISBN 9781496242761 (epub)
ISBN 9781496242778 (pdf)
Subjects: LCSH: Starlings—North America—History.
Classification: LCC QL696.P278 S73 2025 | DDC 598.8/63—dc23/eng/20240514
LC record available at https://lccn.loc.gov/2024022118

Designed and set in Questa by K. Andresen.

For all who flock in search of belonging.

Contents

List of Illustrations   xi
1. The Bird Man Cometh   1
2. Mr. Schieffelin's Birds   9
3. A Frenzy of Introductions   17
4. The Sparrows   25
5. Across the Sea in Cages   33
6. Lessons from Down Under   41
7. Occupation   47
8. European Origins   57
9. The Skies Transformed   65
10. Appetites   73
11. Sing a Song of Starlings   81
12. Under Siege   91
13. In Defense of Starlings   101
14. How to Kill a Starling   109
15. Blast 'Em with Starling Calls   119
16. Darkness in the Golden State   127
17. Rise of the Bird Men   135
18. Death from Above   147

19. Can't Beat 'Em? Eat 'Em 157
20. Mapping the Travelers 165
21. Poison Years 175
22. A Forever War 185
23. Spellbound 195
24. Built for Survival 205
Epilogue 211
Acknowledgments 221
Notes 223

Illustrations

Following page 100

Perched starling
Eugene Schieffelin portrait
John Gould lithograph
Bison and starlings
Song of the starlings
Song of the starlings 2
C. K. Berryman cartoon
Rachel Carson portrait
Otto Standke flyer
Record of starling distress calls
Electra crash
Large flock of starlings at a cattle feedlot
Starling droppings on a trailer
Starlings in a tree
Bird bands
Starling murmuration
Starling wing
Josiah Whymper engraving of a starling

STARLINGS

1

The Bird Man Cometh

HOPE RODE INTO MOUNT VERNON, NEW YORK, IN THE HUMID summer of 1959 with the arrival of the Bird Man. After a long train trip from Kansas, he checked into the $7-a-night Hartley Hotel downtown and unpacked the tools of his trade, among them some aluminum paddles, a noisy chime worn around his neck, and a gray double-padlocked metal box about two feet long. No one but the Bird Man was allowed to look inside the box, lest anyone spy the secret to completing the job he'd been hired to do: drive away the ten thousand or so European starlings infesting the city's trees, parks, and neighborhoods.

The birds, with their din and copious droppings fouling sidewalks and houses, had troubled the city for five decades, especially in the summer, and the city fathers had had enough. Five years earlier, desperate crews had mounted a loudspeaker in town that blasted the recorded shrieks of an injured starling. The noise drove the starlings away but not for long. In August 1959 the city signed a contract to pay the Bird Man $4,000 to send the starlings on their way for good. He claimed he'd done it in Louisville, Indianapolis, Wichita, and back home in Great Bend, Kansas. Where others had found only failure and disappointment shooing starlings around the country, it seemed the Bird Man had somehow found a way.

"I do it with a secret method that I ain't going to talk about," he told curious reporters in his high-pitched twang after arriving in Mount Vernon.

People ask me to chase birds and I chase 'em, but I didn't come 2,000 miles to tell you how it's done. You don't think a man who's as old as I am and has a secret worth half a million is going to blab it all away, do you? No, sir, not The Bird Man. I don't hurt them, but when I chase starlings, they stay chased. I can drive 'em out of one tree and into another if I want to. I can drive 'em out of Cleveland and into Cincinnati if I want to. I can do anything with 'em because I know all about 'em, that's my secret.[1]

The enigmatic hero, described by one reporter as a "vigorous, wizened little man," was Otto D. Standke. He was seventy-one, bespectacled, and fond of having a cigar in his mouth. Before he got into bird chasing, he'd sold phonographs for Columbia and worked as a salesman and furniture buyer for Montgomery Ward. In Mount Vernon, he wore a pin on his lapel claiming that he'd made more than $1 million in sales during his career. On his tie was an engraved silver clip that read "The Bird Man."[2]

Standke had stumbled onto his latest enterprise a decade earlier. Every summer thousands of starlings roosted in the elms of a city park in Great Bend. The birds had taken over—even band concerts were canceled. City officials had tried to scare them away by planting aluminum owls in the park. "The starlings liked 'em so much they took to roosting on their heads," Standke said. Frustrated that the city had spent $1,500 on the failing fake owls, Standke had taken matters into his own hands. One night he sneaked down to the park's bandstand with a flashlight and caught twenty-four starlings by hand, according to the story he told. He spirited them back to his pheasant and turkey farm and released them into a barn. He spent the next two months studying their every move, eschewing "tomfool books" and learning firsthand how starlings worked, he said. The next summer, he went to the park and cleared out the birds using his own still-confidential method for chasing them away. "They haven't been back since," he bragged in Mount Vernon.[3]

The Bird Man couldn't get to work soon enough, as far as the Mount Vernon locals were concerned. The starlings had been a terrible fixture for far too long. Thirty years earlier, just after dawn on an October morning in 1929, thousands of starlings and blackbirds had descended on the city, hungry and obnoxious. Most ate their fill and moved on; hundreds more died on the spot, exhausted.

"Patrolman Charles Schulz said that while walking along one of the business streets, the skies seemed to rain birds upon him and he took shelter in a doorway," one news account said. Decades later a local recalled when the starlings moved into a quiet, sweet-smelling neighborhood: "Gone suddenly were the peaceful evenings in that New York City suburb, where previously only sweet birdsong and the sound of children at play broke the silence; instead the air was filled with shrill screeches that resembled nothing so much as fingernail scratchings on a blackboard. On any fine evening during the starling invasion, sensible pedestrians carried open umbrellas and stepped cautiously."[4]

The situation was no better a few miles away in the heart of New York City. Starlings once famously roosted by the thousands at the Metropolitan Museum of Art. But by the spring of 1959 the biggest flocks—enormous, twisting sheets of birds in the sky numbering fifty thousand to a hundred thousand—had taken up residence on the steel girders supporting the Riverside Drive Viaduct. They arrived in the evening in dark clouds of wings and feathers, produced an all-night racket in the neighborhood, and slipped away at dawn. They had a reputation as being noisy, filthy, and utterly unlovable, occasionally causing a stir when they collided with the girders and suddenly, surreally, fell dead from the sky.[5]

Whether the people of Mount Vernon and New York City knew it or not, they were living a shared experience with people across the continent, urban and rural alike. After European starlings were introduced just a few miles away in Central Park in the late 1800s, they had marched across America like a conquering army. They were ravenous, loud, tenacious, and quick to breed. Hundreds became thousands became millions and, in time, hundreds of millions. In the air, great flocks swirled in mesmerizing clouds like giant ribbons of black smoke writhing above the horizon in a rhythm only they seemed to understand. On the ground, the effect was much more sobering. Starlings could descend on a crop and eat a year's worth of effort in a day or two. City trees swayed under the smothering weight of roosting starlings, and buildings and sidewalks were coated with excrement. Sometimes they stayed for weeks or months or even longer. They seemed as fickle as they were fearless. A nuisance at best, they bordered on apocalyptic when viewed through the darkest lens.

When it came to dealing with the local native birds, the newcomers seemed to rely not on violence but bother and perseverance, said one ornithological report not long after starlings' arrival in America. "Most of its battles are won by dogged persistence in annoying its victim than by bold aggression, and its irritating tactics are sometimes carried to such a point that it seems almost as if the bird were actuated more by a morbid pleasure of annoying its neighbors than by any necessity arising from a scarcity of nesting sites," the report said.[6]

Within decades of being introduced, starlings were among the most numerous birds in America and also one of the most hated. Indeed, famed biologist Rachel Carson once quipped that the starling "probably has fewer friends than almost any other creature that wears feathers." Later, another writer said, "Francis of Assisi, if he ever tangled with them, might have been tempted to whittle himself a slingshot."[7]

As the starlings proliferated, people attempted in vain to keep them away using creative, audacious, and occasionally absurd techniques, including nets, itching powder, live owls, fake owls, mechanical hawks, firecrackers, and flashing lights. They tried draping buildings with electrified wires, oiling ledges, blasting recorded noises from speakers, and even strewing dirty underwear in a tree. Countless men armed themselves with fire hoses or shotguns, and cash bounties were offered. The military was called in, as were the top minds in wildlife biology. Chemical poisons were developed specifically to kill starlings, airline engines had to be redesigned after a horrific crash in Boston, and crews were hired every fourth January in Washington DC to keep them out of the trees lining the streets of the presidential inauguration parade route. More experiments were conducted, better ways of killing devised, reports written, curses cast, congressional inquiries made. Through it all, exasperation multiplied as starlings moved, expanded in range and numbers, and persisted, mysterious as always in their fluid flocks, stubborn as stone, reliable as the tides.

Still, when the Bird Man arrived in downtown Mount Vernon in the sticky, steamy August of 1959, he liked his odds. At twilight on the first night, Standke collected his tools, pulled on a red plaid cap, and made his way to Commonwealth Avenue, the heart of a stately six-block neighborhood where the starlings had stubbornly roosted. Word of his arrival preceded him. The fire department roped off the street, and hundreds of people gathered to watch, including children in their pajamas. With some showmanship, the Bird Man hung the two-foot-long chime around his neck—essentially a metal pipe on a rope—and put long metal paddles on his hands. "Puffing on a cigar, he marched off briskly, banging the flappers together once every second, clanging the chime once every five seconds," according to one account.[8]

The well-read among the bystanders might have been reminded of the "Labors of Hercules," the Greek myth in which the hero has to perform twelve tasks as penance following the death of his wife and children. The sixth required Hercules to drive away an enormous flock of birds—sometimes described as man-eating birds—that gathered at a wooded lake near the town of Stymphalus. He was unsure how to proceed until the goddess Athena arrived with a pair of brass *krotala*, noise-making clappers similar to castanets. Climbing a mountain, he rapped the devices, flushed the birds, and triumphantly shot them with a bow and arrow (or perhaps a slingshot).

But Otto Standke was no Hercules. At first the starlings just fell silent. Standke carried on like this for forty-five minutes, walking up and down the street raising a racket, while the crowd stood by. Most of the birds eventually picked up and left, flying out of the neighborhood and into the night. The mayor was impressed, saying that if the birds were still gone in a month, Standke would get $1,000. If they were still gone in three months, he could count on another $1,000 payment. If the city remained starling-free for a year, something that seemed akin to a suburban miracle, the Bird Man would get his final $2,000 check.

The *New York Times* sent a reporter and photographer out to Mount Vernon the next evening. They shadowed Standke as he moved into the Chester Park area, marching down the street "like a one-man band" and sometimes standing at the base of a tree, clanging until starlings took flight, only to land a block or two away. "Mount Vernon officials concede that his methods were a bit bizarre but they do not care if Mr. Standke uses alchemy, witchcraft or moral suasion," the *Times* noted. "If he gets rid of the starlings, he will get his money."[9]

In the days that followed, more reporters came out to cover the story of the Bird Man from Kansas. He soaked up the attention and coyly teased about his mysterious double-locked box, allowing a reporter to shake it. The journalist proclaimed that it sounded like nothing more than a fist-size rock in some dirt. Standke refused to say more about his methods as reporters followed him around, beguiled and amused. "After three nights of this procedure, neither the starlings . . . nor the people of Mount Vernon could make out whether The Bird Man was a wizard, a spellbinder or an outright charlatan," *Sports Illustrated* said in a story that week headlined "Otto and the Night Visitors." "But one thing seemed clear. Faced with the continuing nuisance of the pesky, defiling birds, the solace-seeking suburbanites of a neurotic century are willing to try almost anything—or anybody—that offers them peace."[10]

The spectacle continued for several more nights. The Bird Man clamored with his paddles and chime. The starlings fussed and moved around but refused to leave the city. People came out with their own noisemakers, including garbage can lids, hoping to aid the dispersal. Standke wouldn't have it. He stalked out of Chester Park, claiming it wouldn't work having so many people getting in on the act. He was still sulking in his hotel room a couple of nights later as his seven-day deadline expired. He bristled at a report from the city health inspector that there seemed to be more starlings than before the Bird Man's arrival.

"I don't like that doctor," Standke told reporters. As the final hours of his contract ticked away, Standke wouldn't budge. "I'm not going out tonight."[11]

The Bird Man, defeated, checked out of the hotel soon afterward, claiming city officials in Youngstown, Ohio, wanted his help to clear starlings off some government buildings so he'd be moving on. Standke never collected his $4,000. Accounts differ, but it appears he was paid just $245, or possibly $400, for his troubles in Mount Vernon.

When persuasion and harassment failed with the starlings of Mount Vernon, the city tried brute force. The following summer, when the starlings inevitably returned, city officials instructed the police to simply fire shotguns at the birds. Forty were killed in one night; the rest took off but soon returned. A news story declared, "The shooting will be resumed each evening the weather permits."[12]

2

Mr. Schieffelin's Birds

WHEN EUGENE SCHIEFFELIN WOKE UP IN MANHATTAN ON A Tuesday morning in April 1889, it was cool outside, but the temperature warmed into the upper fifties by the afternoon. Spring was settling in after a long, cold New York winter, and the city was, as ever, alive with the frantic buzz of newcomers and old families alike trying to find a way ahead.

Schieffelin was sixty-two, married, the seventh son of a prominent lawyer and businessman, and part of a storied, complicated family that had been in New York for nearly a century, building its fortunes in the pharmacy business but also through trade, retail, and real estate deals. His roots went back to what is today southern Germany and Switzerland. The first of the family to come to the continent, Jacob Schieffelin, arrived in Philadelphia in 1740 but soon returned across the Atlantic. His son, also named Jacob, came to Philadelphia in 1845, married, stayed, and gave birth to a son, yet another Jacob, in 1757. It was the third Jacob Schieffelin, Eugene's grandfather, who started the family drug business in New York. In the ensuing generations, the enterprise survived fires, cholera outbreaks, financial panics, the seizure of its ships in Amsterdam on the orders of Napoleon, and the Civil War, and it saw the rise of railroads, the arrival of the telegraph, and the miracle of electricity.

In Manhattan, the family had warehouses and storefronts, a cadre of loyal employees, and an entrepreneurial spirit, often selling more than just the medicines they'd had shipped from overseas. Their diversification included the sale of coffee, sugar, and gunpowder. They were also among the first to sell gasoline in the city. They pursued land deals, some that benefited themselves and others the community, including donation of the land where St. Mary's Episcopal Church would be built in what was then called Manhattanville. The family proliferated in New York, and at some point a street, Schieffelin Avenue, was named for them in the Bronx.

But pharmaceuticals, sold wholesale and retail, remained at the heart of the Schieffelins' enterprise. In the early 1880s they built a laboratory to develop new drugs and research better ways to manufacture them. Pharmaceuticals were still often regarded with skepticism at that time, and the Schieffelins endeavored to infuse them with an air of respectability. The family business, one self-produced family history said, was part of a long-term undertaking to divest the drug business "of the mysteries and superstitions in which it had been enshrouded by alchemy." Through it all, the company held the Schieffelin name and was run by one or more of its sons. Eugene's brothers had been at the helm from 1849 until the end of the Civil War, and then the next generation took over.[1]

The Schieffelins, well known and well-to-do as their business blossomed in the nineteenth century, rubbed shoulders with many of the movers and shakers in Manhattan and beyond. They turned up in the newspaper society columns, had a spacious family home on Madison Avenue, and held memberships in many of the city's upper-crust clubs.

For his part, Eugene seemed to go along as expected. He occasionally worked in the family pharmacy business but never had his name at the top of the masthead. "He had little inclination . . . for active business pursuits, but was known for rare intellectual qualities, the result of inherited tastes and talents, as well as of careful study and cultivation in literature, the fine arts and sciences. He was possessed of much accomplishment of manners and address, and of unusual conversational gifts," said one biographical sketch from 1907, adding, "He was greatly interested and deeply learned in ornithology."[2]

Sometimes historic records mention that Schieffelin was a druggist; other times, perhaps reflecting his place in society, they refer to him simply as a gentleman. On one passport application, he listed his occupation as "man of leisure." He and his wife, Catherine, had servants to take care of them, and without children, Eugene pursued other interests. He was a member of the Saint Nicholas Society, the New York Genealogical and Biographical Society, and the Society of Colonial Wars. He was also an "artist of some distinction," once painting a full-length portrait of Gen. Philip John Schuyler that was later hung in the Saint Nicholas Club.[3]

Had he read the morning papers that day, April 23, 1889, he would have seen the dominant story: The Oklahoma land rush had started the day before, and more than fifty thousand dreamers had flooded in to lay claim to their piece of two million acres of "unassigned land." There was also much chatter about the arrival in Philadelphia of the steamship *Missouri*. It was carrying more than three hundred exhausted passengers and crew members from the ship *Danmark*, which had wrecked in the Azores a couple of weeks earlier, prompting a dramatic rescue at sea.

But it's likely Schieffelin's mind was on other things that morning—namely, the seventy-two glossy European starlings in cages that he'd had shipped over from Europe and what would become of them later that day. For decades, he'd been sporadically buying birds from overseas and releasing them in New York and elsewhere, including sparrows, nightingales, and quail. This had become something of a hobby among Schieffelin and his friends. Most of the time, the releases didn't amount to much, and the birds didn't survive in the wild. But even if it wasn't immediately obvious that morning, this release would be different.

History doesn't record exactly how it happened, but what's known is that sometime that day, Schieffelin traveled the few blocks to Central Park, perhaps with some servants from his house, with several cages of the noisy birds in tow. No doubt it created a commotion and a bit of a spectacle. Here was a man, scion of a prominent Manhattan family, now mustachioed and going gray, moving with a mission through the busy streets with a flock of captive birds of a species that hardly anyone on the continent had ever seen. We don't know what Schieffelin made of the looks he got along the way, but it's easy to suspect he went with a sense of pride and purpose. He would have been careful of his investment.

If he'd paid roughly the going price at the time, around $5.50 per pair, the starlings would have been worth around $200. He knew birds were fickle things, so no need to jostle them any more than they'd been tossed about in the weeks-long journey across the ocean. He wanted each starling to live free, thrive, feast in the wild (especially on some of the insects that had become such a bother), and enjoy what their new home had to offer.

After all, the birds had their charms. Starlings didn't hop like most other songbirds. They strode, one foot in front of the other, with a tottering gait. They had a knack for mimicry, too, with voices so tuneful that Mozart even had a pet starling that carried one of his melodies. Their flight patterns, especially in giant flocks, were a sort of mad poetry in the sky. And that plumage. One could get lost in the constellation of stars sprinkled across their glossy black feathers in winter, to say nothing of the iridescent green and purple apparent to anyone willing to take a closer look.

Once at the park, did Schieffelin give a little speech before unlatching the door of each cage? Did he quietly watch as the birds left their holding cells and strutted around on the ground, unsure of their new surroundings? Was he forced to wave his arms to shoo them away, or did they disperse on their own?

The release was little noted in public. One exception was a brief write-up in *Forest and Stream*, a weekly journal in New York focusing on hunting, fishing, and other outdoor activities: "On Tuesday last, Mr. Schieffelin set at liberty in Central Park, this city, seventy-two European starlings (*Sturnus vulgaris*). They at once adapted themselves to their new surroundings, and after taking a bath in the stream flew off to the lawns in search of food."[4]

Schieffelin's starlings stuck around Central Park for several weeks, roosting here and there, before making their way north. It was soon reported that some had made it to Fort Schuyler in the Bronx, on the edge of Long Island Sound. "It is to be hoped," William Conklin, director of the zoo at Central Park, said after Schieffelin's liberation of the starlings, "if they were spared by the bird catchers, that they will return to the Park this coming spring."[5]

The following year, on March 6, 1890, Schieffelin returned to Central Park and released eighty more starlings. This time it made the daily papers. "The starling is little known in America but these eighty beauties ought to make a good starter for a more intimate acquaintance," the *New York Evening World* reported. "The starling is a little larger than the red-winged blackbird, and not so large as our crow blackbird. It is speckled all over its breast and back with white dots. It belongs to the crow family but is one of the sweetest of songsters."[6]

The next spring, on April 25, 1891, Schieffelin turned loose forty more caged starlings in Central Park. These were his last, as far as we know. Schieffelin couldn't have known it at the time, but the starlings he released over those three consecutive years in Central Park formed the basis of one of the largest occupations of a foreign bird ever recorded in North America and certainly the most confounding. Many a starling hater in the ensuing century has muttered his name with some measure of befuddlement and disgust. What in the world was Schieffelin thinking?

ONE OF THE FIRST STORIES I HEARD ABOUT STARLINGS WENT like this: In the late 1800s there was a quirky rich guy in New York City who was obsessed with William Shakespeare. He wanted every bird ever mentioned by the Bard to live in the United States, so he set a bunch loose in Central Park. So if you don't like the millions of starlings living today in North America, blame Shakespeare and this misguided acolyte who had more money than sense. The story—Eugene Schieffelin's folly and this nutty thing about the Bard's birds—hardened into fact long ago and is still the anecdote many people tell when talk turns to starlings.

As daffy as the tale was, some pieces fit together. During Schieffelin's lifetime in Manhattan, there was a concerted push among movers and shakers to revitalize interest in Shakespeare's works, which had been waning among the American public since the Civil War. A Shakespeare monument was dedicated in the spring of 1872 in Central Park. In 1912 came the four-acre Shakespeare Garden, featuring some of the herbs, shrubs, trees, and flowers mentioned in his plays. And the ornithological significance of the Bard had also come up. In 1871 a book was published called *The Birds of Shakespeare: Critically Examined, Explained, and Illustrated*. Across three hundred pages, bird scientist James Edmund Harting provided a recounting of more than sixty species mentioned by Shakespeare, including wrens, skylarks, nightingales, buntings, loons, cuckoos, hawks, eagles, and vultures. The starling shows up just once, in part 1, act 1, scene 3 of *Henry IV*. Hotspur is scheming a way to drive King Henry crazy and devises a plan to ensure the king forever hears the name of Hotspur's brother-in-law, Mortimer:

> *He said he would not ransom Mortimer;*
> *Forbade my tongue to speak of Mortimer;*
> *But I will find him when he lies asleep,*
> *And in his ear I'll holla Mortimer!*
> *Nay,*
> *I'll have a starling shall be taught to speak*
> *Nothing but Mortimer, and give it him,*
> *To keep his anger still in motion.*

Years after starlings were set loose in Central Park and began to spread across the country, a narrative emerged combining Shakespeare, Schieffelin, and his starlings, wherein a wealthy, Bard-loving Manhattanite freed the birds as a sort of homage to the writer's works. The story, often repeated but never really fact-checked, became a tantalizingly simple cautionary tale about tinkering with nature, one where a single act of human foolishness—to satisfy some sort of literary desire, of all things—unleashed a tidal wave of unintended, irreversible, and devastating consequences. It's a tidy bit of folklore about the dangers of introducing foreign species that's been repeated ad infinitum when it comes to starlings in America. The cautionary part of the story still rings true, and the basic facts of Schieffelin's release of the starlings aren't in dispute, but there's nothing to back up the claims that he was somehow inspired by Shakespeare.

The first mention of the Schieffelin-starling-Shakespeare connection appears to be in 1947, when Edwin Way Teale, a well-known naturalist and writer, published a story in *Coronet* magazine called "In Defense of the Pesky Starling." In it, he briefly refers to Schieffelin's "curious hobby" of introducing to America all birds mentioned by Shakespeare but provides no substantiating evidence.[7]

It's a glancing reference—indeed, just a few words—but the kind of juicy detail that has been blindly recounted through the ensuing decades in magazines, newspapers, books, conference lectures, and government publications. For those bothering to check, though, it becomes clear that Teale's mention (which also appears in one of his books) is a sort of dividing line in how history records the starling introduction. Before Teale's words, there was no mention of the Schieffelin-starling-Shakespeare connection. Afterward, it's all over the place.

Several people voiced doubts about the story over the decades, but a more definitive take on the matter didn't emerge until some sleuthing academics published a paper in 2021. In "Shakespeare's Starlings: Literary History and the Fiction of Invasiveness," Lauren Fugate of Carnegie Mellon University and John MacNeill Miller of Allegheny College looked into the Shakespeare theory and put the point of origin at the hand of Edwin Way Teale. But, they said, it's hard to say where his idea came from.[8]

Teale died in 1980. Fugate and Miller went through his folders, notes, and source material in his archives in search of clues but largely came up empty. The first mention of the Shakespeare claim shows up in an early draft of Teale's essay next to speculation that the Shakespeare Garden was getting underway around the same time as Schieffelin's starling releases. "This may have influenced Schieffelin in his plan," Teale wrote. Fugate and Miller suggest Teale may have mistakenly conflated the garden with the starling release, noting that the garden at Central Park didn't open until more than a decade after the birds were set free. The final published version of Teale's essay about the starlings didn't include reference to the garden, "leaving only the image of an eccentric individual and his 'curious hobby,'" they said.[9]

Still, it was enough to ultimately seed a narrative that proved impossible to ignore and too good to bother with running to ground. "Suddenly the success of this ubiquitous species could be understood as the upshot of one man's quirky quest—and as a testament to the power of the most famous poet in English letters," the academics wrote.[10]

Teale's claim remained a little-noticed curiosity in the starling story in subsequent years. It didn't arrive fully into the American consciousness until September 1974, when *Sports Illustrated* ran a fifty-eight-hundred-word story called "A Plague of Starlings." The subheadline put a bright spotlight on the Shakespeare connection: "The pretty plan was to bring the Bard's birds to America. It brought troubles by the flock." The story opened with an anecdote about the starling in *Henry IV* and then repeated the claim that the starlings were set loose by Schieffelin as a direct result of Hotspur's impassioned speech. Schieffelin was described as "an elegant and eccentric figure in New York high society" somehow smitten with the misguided idea of bringing Shakespeare's birds to U.S. soil. "If he could have foreseen the results," the article said, "he might very well have made an exception in the case of the starling."[11]

Until recently, no one bothered to thoroughly debunk the Shakespeare story. Who would want to? It's proven a durable explanation for Schieffelin's behavior and a useful allegory for those who might be tempted to futz with nature without fully exploring the potential consequences. The truth, as usual, is more complicated. Rather than being a lone eccentric preoccupied with Shakespeare, Eugene Schieffelin was actually part of a movement that had been gaining steam long before his starlings ever found their way to Central Park. These were men, loosely organized and scattered around the globe, who felt certain they could improve on what nature had already provided, dreaming of better birds in the sky and sweeter songs in the air.

3

A Frenzy of Introductions

IN THE WINTER OF 1854, AS FRENCH LEADERS PREPARED TO get deeper into the Crimean War with Russia, a different kind of group was forming in Paris, one that called together "friends of human progress." They included prominent biologists, botanists, agronomists, academics, landowners, and others. Among the goals: "the introduction, acclimatization, and domestication of animals, whether useful or ornamental." The Société Zoologique d'Acclimatation, as it was called, had seized on a growing sense that France and the rest of Europe were missing out on the benefits of animals and plants in other countries. In particular, they felt that bringing in new species from afar could provide more food and improve crop production. Europe had already benefited from potatoes from America, grapes and wheat from Asia, and the importation of pheasants, peacocks, silkworms, goats, and donkeys. Certainly, more could be done. "We have given sheep to Australia; why have we not taken in exchange the kangaroo—a most edible and productive creature?" said French zoologist Isidore Geoffroy Saint-Hilaire, the group's president. "New conquests of animals and plants will serve as new sources of wealth."[1]

While the group touted high-minded principles about acclimatization, what they were advocating wasn't necessarily new. For thousands of years, roaming bands of people brought things from home with them, including pets, domesticated livestock, wild animals held in captivity, plants, and seeds. Pigeons and geese were kept in captivity going back to Neolithic times some seven thousand years ago. Ancient Egyptians kept pelicans in enclosures, and the early Romans had quail. Dating back to at least 800 BCE, traveling Norsemen brought cages full of live birds on their sea voyages so they could release them to gauge how close they were to land. If the birds didn't return to the ship, it was presumed they had found a perch on land nearby.

As more and more people moved across the globe, more foreign species found themselves on new lands. Sometimes it was accidental, like hitchhiking rats or noxious seeds inadvertently mixed in with more innocuous ones. Other times it was deliberate. Over the centuries, many of the new species became part of the familiar landscape, and it was easy to forget they'd originated in a land far away.

But the meeting in Paris put an emerging practice into vogue, especially among the elite: organized large-scale importation of foreign animals. Within a few years, the French Société was handing out prizes for those who had successfully domesticated alpacas from South America, quaggas from South Africa, and emus and kangaroos from Australia. Napoleon was commended for introducing rock partridges. The king of Württemberg, once a German state, agreed to experiment with angora goats, and Emperor Dom Pedro I of Brazil vowed to try to domesticate camels in his country.

In January 1859 at the London Tavern in Great Britain, a meal was served that featured large pike, American partridges, and African eland, a kind of antelope. The meal's sponsors noted that all those nonnative animals could live in Britain if they were sufficiently desired. Soon the Society for the Acclimatisation of Animals, Birds, Fishes, Insects and Vegetables within the United Kingdom was formed. Other acclimatization societies sprang up in New Zealand and Australia. Each put its own twist on the idea, with some focusing more on keeping animals in cages for the public to view and enjoy—menageries that eventually morphed into modern-day zoos—and others seeking animals to release into the wild in hopes of helping their crops or dealing with pests. In some cases, it was simple sentimentality: immigrants who'd settled in a new country yearned for the tastes, sounds, and quarries of home and gave in to those cravings by importing familiar plants and animals. Little thought was given to whether they'd survive and even less to what they might do to native wildlife that had existed there for eons.

Texas A&M professor Thomas R. Dunlap has said the idea drew on a classic Victorian notion that "nature was not only the most productive but also most beautiful when perfected by human labor." Europeans in North American often had their own idea about creating a kind of paradise in the new land, as did English colonists in New Zealand and Australia. "Human power—or the power of wishful thinking—was strong," Dunlap said.[2] And for those settling far from their native lands, what better way to manifest the joys of home than the birds of childhood?

ON APRIL 20, 1871, THE NEW YORK LEGISLATURE APPROVED the incorporation of a group called the American Acclimatization Society. Based in New York City, the group said its purpose was "the introduction and acclimatization of such foreign varieties of the animal and vegetable kingdom as may be useful or interesting." As it happened, Eugene Schieffelin was named president of the society. Among the other officers were men from some of the city's top families, including Robert Roosevelt, a former congressman and conservationist and current New York fish commissioner (as well as uncle to future president Theodore Roosevelt), and Edward Schell, president of a Manhattan bank and brother of a congressman.[3]

The written record of the society is scant, mostly limited to a few newspaper mentions, so we don't know how often they met or what they talked about. But it's safe to assume they felt a kinship to many of the other acclimatization societies in Europe and elsewhere and wanted to do their part to decorate the landscape with foreign species. They began making their presence known in 1877, buying "a lot of English quail" and some English hares to be set loose in Pike County, Pennsylvania. They also purchased some "English starlings and Japanese nightingales," which they planned to release in Central Park.[4]

By then Central Park's wildlife was already a much-discussed attraction, and the American Acclimatization Society wanted to add its mark. The menagerie there had started small in 1860, just two years after the park's construction began, with a few white swans from Hamburg, Germany, and London. State legislation the next year provided money to set aside sixty acres of Central Park to house more animals. A monkey was added a couple of years later, and soon animals were arriving from around the world: rhinos, hippos, leopards, jaguars, hyenas and a chimpanzee.

The attraction wasn't just about caged animals. In 1864 park commissioners imported fifty pairs of house sparrows from England and set them free. The little brown birds multiplied quickly, but the same couldn't be said for subsequently introduced English chaffinches, blackbirds, and Java sparrows. "Unfortunately, their numbers were so small the birds were lost sight of," said one report. Fifty pairs of English skylarks were also released to the park, only to relocate across the East River.[5]

By June 1877, when the American Acclimatization Society released starlings and nightingales in the park, *Forest and Stream* was bullish on the experiment, proclaiming that the city of New York "is greatly indebted" to Schieffelin's group. The magazine didn't specify how many starlings were released but said it was a large number and that readers should be enthused about these new birds from Europe: "From the moment of leaving the nest it begins to manifest its bright and joyous disposition by singing merrily all day, no matter how inclement the weather, how scanty its supply of food, teaching us a lesson of contentment more effectually than could some of our greatest philosophers."[6]

William Conklin, director of Central Park's zoo, offered glowing predictions for the starlings at the time, claiming they would clear gardens and fields of snails, worms, grasshoppers, and caterpillars. In fact, he pointed out, he'd heard that a single young starling could eat 140 worms or insects in fourteen hours. Another fantastical estimate came from Germany that 180,000 starlings had feasted on 12.6 million snails and worms each day. "We believe that the starlings are vastly more useful as worm destroyers than the sparrows, which have certainly not made a good record for themselves since their introduction here," Conklin said. "They are vastly more entertaining and interesting in themselves, besides being more civil to other birds." Conklin was an early, if sometimes hyperbolic, evangelist for the starlings, noting that they were smart enough to learn to speak and sing in captivity. "One has been known to repeat the Lord's Prayer without missing a word," he boasted.[7]

Conklin showed up at the American Acclimatization Society's meeting in November 1877, saying he had high hopes for the starlings that had been released in Central Park that summer. He anticipated that more birds could be introduced to the park, including skylarks, blackbirds, the "English titmouse," and the chaffinch. The payoff, though, would have to wait.

IN THE YEARS THAT FOLLOWED, THE STARLINGS THAT EUGENE Schieffelin and his friends had let loose in 1877 were nowhere to be found. They'd either flown away to some distant hidden territory or quietly died out just beyond human view. It wasn't uncommon for introduced birds, despite the hopes of their eager patrons, to simply vanish after fleeing their cages. Adaptation to a new environment could be tough. The birds faced unfamiliar kinds of weather, uncertain prospects for food, and disorientation in the new world around them, and they had no one to show them the ropes. Still, it's easy to imagine Schieffelin venturing often into Central Park, peering into the trees and listening for any signs of his birds, and returning home disappointed day after day. The American Acclimatization Society fizzled out, too, vanishing from the newspapers and no longer being a driving force in importing animals from faraway lands.

Central Park barely noticed. Its branches, lawns, and waterways remained a wonderland of native birds, especially in the spring and summer, including cedar waxwings, thrushes, hawks, flickers, woodpeckers, and finches. "Let no one who wants to study birds despair because he is confined to the city," the *New York Daily Tribune* wrote. "Let him go to the Park and look and listen."[8]

Schieffelin, unsatisfied and stubborn, remained undeterred in his quest to add starlings to the city's list of local birds. More than a decade later he set loose nearly two hundred starlings in Central Park over the course of three spring seasons—1889, 1890, and 1891—and kept a keen eye out for survivors. It wasn't long before a taxidermist at the American Museum of Natural History, just across the street from the park, noticed a few starlings huddling in the eaves of the building. Once Schieffelin caught wind of that development, he made regular visits to the museum's bird department to ask if any more starlings had been seen recently. "He was particularly pleased when he learned that a pair had nested in the building," said one ornithologist who worked there at the time. The pair stayed for three years. Others spread deeper into the city, and within a few years they were in Brooklyn and on Long Island. About twenty starlings were seen on Staten Island in 1891. By the fall of 1893 a flock of about fifty was regularly seen in the Kingsbridge neighborhood in the Bronx.[9]

The stories flowed from there. A rock-throwing boy killed a starling in Brooklyn, and two more were shot in the same neighborhood around the same time. Naturalist George Bird Grinnell was riding his horse on Riverside Drive and spotted a bird he'd never seen before. "The bird was black, with a white bill and a short tail," he said. "It was a European starling, unless I am very much mistaken. . . . I shall endeavor to investigate the matter further."[10]

By 1895 starlings had earned their own write-up in the *Handbook of Birds of Eastern North America*, with their iridescent and spotted summer and winter plumage patterns detailed, range outlined, and behavior noted. "Starlings are *walkers*, not hoppers," it emphasized. "These birds are resident throughout the year, and, as they have already endured our most severe winters, we may doubtless regard the species as thoroughly naturalized."[11]

In 1898 a flock of about thirty starlings was reported north of the city of Sing Sing (today's Ossining) along the Hudson River. Around that time, Schieffelin and his wife, Catherine, along with a few servants, were spending at least part of the year on a farm in Red Hook, New York. They would have passed through this city on their way north, and Schieffelin no doubt scanned the sky and the farm fields for his beloved starlings, swelling with pride if a flock came into view. If he hadn't known it before, it was becoming clear now: his experiment, at least in terms of the starlings he had set free a decade earlier, seemed to be working. Starlings were quick studies of their new environment and adept at pioneering into new territories.

THERE'S NO RECORD OF EXACTLY WHERE IN CENTRAL PARK Schieffelin released his starlings, but I went looking anyway during a trip to New York City on an unusually warm October day. The park was crowded, even though it was the middle of the week. Families cruised by on bicycles, couples lounged on the grass, and photographers posed brides among the fall colors while rattling off instructions in hopes of capturing the last light. Just beyond the green, skyscrapers towered over the park like eager soldiers, and buses exhaled in the stop-and-grow traffic. At one spot, near the U-shaped water feature on the south end known as the Pond, a class of tango dancers stepped to the music coming from an old boom box. Penetrating the music was a cacophony of cries and high-pitched caws from hundreds of starlings perched in the trees overhead. The bird calls, tangled and out of rhythm, pulsated but barely seemed to register with any of the people below. I remembered the description by an English ornithologist a century earlier who noticed that starlings sometimes seemed to sing "a special note" when flying to their roost later in the day or departing in the morning. "It is low, of a musical quality, and has in it a rapid rise and fall—an undulatory sound one might call it."[12]

At one point, a few starlings fled a branch, followed seconds later by most of the flock. They were changing trees, moving just a few feet to the west, only to settle down, cry out, and move back again. As fussy as they seemed, I had to assume they had their reasons.

It's not hard to understand why Schieffelin brought his starlings here. There are more than twenty thousand trees in the 843-acre park, an ideal refuge for resident and migrating birds alike. More than two hundred bird species have been recorded in Central Park, from sharp-shinned hawks and peregrine falcons to delicate cedar waxwings and fish-hungry cormorants. It's a haven for binoculared birders, too, drawn to the waterfalls and pools of the North Woods and the lush expanses of the Ramble, where the tree canopy is sometimes so thick it blocks out the sky.

The south end—with its rocky outcroppings, water features, rolling terrain, and plentiful trees—seemed like a potential spot for Schieffelin's starlings. It would have been close to the zoo and not far from his house. I watched the flock that had gathered in the limbs overhead, squinting for a leader who might guide them from branch to branch, and wondered how many of them could trace their lineage back to the batches that Eugene Schieffelin had brought to the park more than a century before. How many generations was it now? Somewhere along the line, they proliferated, expanded, and became deeply despised. A narrative had coalesced around them that they were an ugly invader, an unwanted pest that had brought little more than trouble and filth. None of them asked to be here, and yet here they were, prattling on high in the trees in a land that none of their distant ancestors had known. I wondered if birders ever trained their scopes on the starlings and took any delight in their presence here. Or were they simply ignored as pesky outsiders taking up space where the "real" birds were supposed to be? And come to think of it, how long must starlings be here before they're finally considered one of the locals, as Schieffelin and others had hoped?

    A few years back, a National Audubon Society blog post had inveighed against starlings, giving birdwatchers permission to scorn those in the avian world deemed too damaging to people or fellow birds. "It's okay to hate certain species, too—healthy even," the writer said. "I suggest you start with European starlings."[13]

4

The Sparrows

BEFORE STARLINGS, THERE WERE HOUSE SPARROWS, ALSO known as English sparrows.

In the middle of the nineteenth century, inchworms beset New York City. Copious numbers of the little devils, squirming larval forms of white linden moths, devoured leaves and shrubs and found their way down shirt collars and into people's hair. Infernal pests, they had a seemingly fiendish ability to unsettle and unnerve. "Branches teemed with fat, leaf-bloated caterpillars, which hung at the ends of long threads, brushing against the cheeks of hapless pedestrians, dangling over hat brims, slipping down necks, and crawling up backs and sleeves," went one description. "Others fell directly to the sidewalks, where passing feet mashed them into a sickening mush."[1]

Tar, turpentine, petroleum, nothing seemed to keep them away. "Worms rolled off awnings, buildings, trees and a dozen other things upon unfortunate passersby," one newspaper recalled. "They infested everything. No person knew whether he ate them or not, especially if it was his habit to dine at a restaurant."[2]

The story goes that a sea captain came to the office of a Brooklyn dentist, Dr. Salmon Skinner, and found an inchworm in the chair. He mentioned that such a thing wouldn't happen in London because the house sparrows were such voracious consumers of inchworms. The next year, 1848, at Skinner's request, the captain brought a crate with a hundred of the sparrows to New York. Skinner and the captain released half of them near Skinner's home at the corner of Montague and Henry Streets and the rest at a vacant area that eventually became Madison Square Garden. The birds didn't survive long in their new home, and the worms soldiered on.[3] Still, the idea of house sparrows as heroic worm eaters had been planted.

In addition to pulling rotten teeth, Skinner was a member of the Brooklyn Institute of Arts and Sciences, which decided to pay for the importation of eight pairs of these sparrows to the city in 1850. The birds spent the winter in a cage in New York City but died soon after they were released. Undeterred, the committee put up $200 and sent Nicolas Pike, director of the Brooklyn Institute, to England. There's some dispute in the historical record about whether this happened in 1852 or a few years later, but whatever the case, in Liverpool that autumn Pike put in an order for a large number of sparrows and other songbirds. "They were shipped on board the steamship *Europa*, if I am not mistaken, in charge of the officer of the ship," Pike recalled later. The birds eventually were taken to the home of another member of the committee, John Hooper, who held them for the winter. In the spring they were let loose on the grounds of Greenwood Cemetery. "A man was hired to watch them," Pike said. "They did well and multiplied."[4]

News of the sparrows' success in New York spread quickly, and more were bought and set free, including in Portland, Maine, in 1854 and Peacedale, Rhode Island, in 1858. They were in Quebec by 1864 and in Galveston, Texas, by 1867, and soon do-gooders were setting them free in San Francisco, Utah, Michigan, and Wisconsin. A thousand sparrows were let go in Philadelphia. It was a legitimate craze. Years later a government ornithologist tried to tally up all the different instances where sparrows had been released—everything from a few pairs in St. Louis to several hundred in Indianapolis—and determined that it was more than a hundred and that his estimate was likely only a small fraction. "They were welcomed everywhere. Clubs were formed and the sparrow became a status symbol," one report said later.[5]

Back in the tony parts of Manhattan, where worms devoured the leaves of the shade trees, Eugene Schieffelin imported fifty pairs of sparrows and set them free in Madison Square near the family home in 1860. He did the same in several successive years, at least until 1864. Within a few years, the worms were mostly gone, and many locals seemed pleased at the sparrows' service.[6]

A noted New York ornithologist named George. N. Lawrence spotted some of the "gentle and fearless" sparrows in Manhattan nestled in the ivy on the walls of a church at the corner of Fifth Avenue and Twenty-Ninth Street. "I afterwards learned from our associate, Mr. Eugene Schieffelin, that he had been looking after them with much interest; in fact he is entitled to the credit, in a great measure, for this important acquisition to our city."[7]

In the fall of 1868 the *New York Times* ran a long piece extolling "our feathered friends" the sparrows and their grand appetite for the city's inchworms, even raising hopes that they might turn their attention to other undesirable insects. "Let them come," the paper said of the sparrow, "the more the merrier." The *Times*, though, noted how fast the sparrows had spread through the city and that a time might soon come when there would be plenty, if not a bit more than needed.[8]

The sparrows fanned out across the continent, sometimes even hitching a ride in boxcars hauling corn. Ever the colonizers, they always found a way to move, eat, breed, and continue on their way. "The bird occupied the continent with considerably less effort than its human counterparts," one writer said in a history of the house sparrow's expansion.[9]

Indeed, they seemed the perfect survivors for the Americas. House sparrows are master adapters, capable of foraging for a variety of foods, including bugs, crops, and leftover human food, and expanding rapidly. They nested in the holes of buildings, on top of streetlights, in barns, and even in vines that scaled walls. They were also copious breeders, with a pair in New York City typically capable of producing twenty to thirty young per year. As a mathematical exercise, Walter Barrow, an ornithologist at the U.S. Department of Agriculture, calculated that a single pair of these sparrows—assuming their descendants all stayed close to home and produced an even mix of males and females—could be responsible for more than 275 billion offspring within a decade, a rate of expansion "without parallel in the history of any bird." "Like a noxious weed transplanted to a fertile soil," Barrow wrote, "it has taken root and become disseminated over half the continent before the significance of its presence has come to be understood."[10]

By 1870 house sparrows were established as far south as Columbia, South Carolina, and as far west as Davenport, Iowa. Later that year the *New York Times* changed its tune about sparrows, suddenly lamenting their vast arrivals in suburbs, the deleterious effects on native birds, and the messes they left behind. But what to do? "It would, however, seem cruel to invite these little chaps over to our boasted land of liberty only in due time to transmute them into the tender broil or the savory pot-pie," the *Times* observed, "and let us hope that that day of dire necessity may be long averted."[11]

Government scientists estimated that the sparrows had gone from occupying about 15,000 square miles in 1880 to more than 885,000 square miles six years later, with homes in at least thirty-five states. More than forty million were thought to be living in Ohio alone. No one knew for sure, but it was estimated in 1884 that there were an astonishing nine hundred million house sparrows in the United States.[12]

Thus for the first time in North America's history, house sparrows had become a common sight for almost anyone interested enough to look. Hopping around on spindly legs and congregating in noisy groups, they squabbled over food morsels and battled for territory. Hopeful males puffed up their chests and opened their wings in a never-ending quest for female companionship.

Any benign fascination with these sparrows, at least among most people, was short-lived. In Philadelphia, the inchworm was gone within about five years of the sparrows' arrival, only to be replaced by a hairy caterpillar that the sparrows wouldn't eat. So the little birds went looking elsewhere. Soon the sparrows overran croplands, eating through farmers' fruits, grains, and vegetables. They were expert and aggressive colonizers, soiling cities, congregating in neighborhoods with their incessant cheeping, and driving out native birds, sometimes attacking them. "While the worms had been bad," one person lamented, "the sparrows were worse. It became apparent that something had to be done that would free us from this tiny Frankenstein we had created."[13]

Some felt betrayed by the introduction, feeling that it had been a moneymaking scheme by the bird sellers, who had failed to deliver on their flowery promises. "If there is one single redeeming quality possessed by these unmitigated nuisances, the English sparrows, we do not know what it is," a Virginia letter writer complained in *Forest and Stream* in the summer of 1881.[14]

The editors of the magazine also took a swipe at the sparrow introduction in the same issue. First they ran a celebratory poem from William Cullen Bryant written in 1858, extolling the arrival of this "stranger bird" and its insect-killing ways, with the last stanza reading:

> *And the army-worm and the Hessian fly*
> *And the dreaded canker-worm shall die;*
> *And the thrip and slug and fruit-moth seek*
> *In vain to escape that busy beak;*
> *And fairer harvest shall crown the year;*
> *For the Old-World sparrow at last is here.*

The editors noted that Bryant had written his ode after visiting W. H. Schieffelin in Manhattan, reporting that Schieffelin had bought a number of sparrows and released them in his Madison Avenue garden. William H. Schieffelin was Eugene's nephew, who soon took over the family business and later got involved in a movement to bring more foreign birds to the United States. Right below Bryant's poem, *Forest and Stream* editor Fred Mather ran his own six-stanza satire that railed against the sparrows for driving away native birds and generally being a pain. One stanza went like this:

> *He defiles our porches, there's no denying that;*
> *He has ruined my wife's dresses and spoiled my best hat.*
> *He hangs round the bird cage to pilfer the seed,*
> *And gives the canary a foul insect breed.*
> *He never eats worms, let us tell it abroad,*
> *This Old World sparrow is a terrible fraud.*[15]

The American Union of Ornithologists offered its own scathing indictment of sparrows in 1884, with the *New York Times* paraphrasing the group's stance like this: "The sparrow is an impostor, a thief, and a murderer, and recommends that he be exterminated without any further delay. Filled with hatred of all honest birds, the sparrow makes war upon our native song birds, and according to the testimony taken by the committee, he is rapidly exterminating them. Meanwhile, he pays no attention to worms, and thus entirely ignores the contract under which he was brought to this country. He is a faithless, graminivorous, murderous thief, and even the most hardened crow looks upon him with contempt and disgust."[16]

By the 1880s states started offering bounties for dead house sparrows. Michigan paid a cent for each one and then bumped it to 3 cents apiece. Utah had to keep raising the amount of its bounty, first from less than a penny per sparrow to around 3 cents, and paid a nickel for every dozen sparrow eggs that were destroyed. Ohio and Illinois were forced to take similar steps. These actions didn't come cheap: Michigan and Illinois combined paid out more than $117,000 in bounties in less than a decade. Not that the results amounted to much—the bounties were like throwing a thimble of water on a raging wildfire. "They have not exterminated the English sparrow or even caused a perceptible diminution in its numbers except in a few localities," said a government report in 1899.[17]

In the "Sparrow War," as it was sometimes called, people fell into two camps: those who welcomed the house sparrows to America, for their songs or cuteness or appetite for insects, and those who didn't, for fear of the havoc they'd wreak on their new land. Even then, the unintended consequences of "biological control" were understood in many circles. The sparrows were so controversial, and the situation had become so complicated and widespread, that the birds were the subject of the first-ever bulletin by the U.S. Department of Agriculture. The report, published in 1889, filled a remarkable four hundred pages. The primary author was Walter Barrow, the assistant ornithologist at the department who'd made the earlier calculations. In this report, he provided a painstaking account of the birds' spread and damage. As a professional scientist relying on data and long-term observations, he struggled to contain his contempt for those who seemed simply pleased to have the sparrows around, consequences be damned.

"It would be folly to expect all friends of the Sparrow to accept our conclusions," Barrow wrote. "There are some persons whose minds are so constituted that nothing is evidence to them except what is derived from their own observation, and as this unfortunate mental infirmity is commonly correlated with the total inability to observe anything which interferes with their theories, it makes little difference whether their opportunities have been good or bad, their position is unassailable."[18]

Still, there was no discounting the allure of the sparrows' "familiar chirp" among Europeans now living in the states, Barrow noted. That was especially true for those living in the cities and those of English, German, and French descent who were "inspired by the collections of the birds of their fatherland." The situation was worsened by the "prevailing ignorance" of most Americans when it came to native birds, combined with the "totally erroneous, or at least grossly exaggerated" benefits of the sparrows.[19]

As news reached across the Atlantic, the English were amazed that the Americans would import such a "worthless bird," one report said. "Anti-sparrow organizations had existed in England since 1744 and almost every parish was spending money to destroy sparrows."[20]

Discussion intensified around finding a solution to the sparrow problem in the United States. There was talk of guns, poisons, and even the potential for unleashing domestic cats to exterminate the little birds. It all seemed impossibly futile, and indeed it was. "Let us accept the inevitable with good grace and cease to abuse the sparrow," the *Times* said in the midst of the 1880s panic over sparrows. "We brought him here, and here he will remain."[21]

Not everyone was ready to relent. The sparrow problem had gotten so bad in Pittsburgh, Pennsylvania, that city officials began breeding a few starlings, imported from Europe, at a city park in hopes of releasing them to drive away the sparrows. "The starlings are of the German variety. Like their own beloved Kaiser, they are very warlike, and this is the reason why they were preferred to denizens of Great Britain. The German starling is a most aggressive bird and sees no reason why he should not assert his supremacy at any and all times," one newspaper wrote in November 1897. The plan was to release forty pairs of starlings the following spring, unleashing "a duel such as never has been known in ornithological history."[22]

The newspaper noted some irony in the starlings being imported to combat another bird that itself had been brought in to fight off a pest. The story reminded readers of this old verse, often attributed to Jonathan Swift or Augustus De Morgan:

Big fleas have lesser fleas
Upon their backs to bite 'em,
And these same fleas have lesser fleas
And so on infinitum.[23]

By 1903 author William Leon Dawson offered his own appalled take in *The Birds of Ohio*: "Without question the most deplorable event in the history of American ornithology was the introduction of the English sparrow."[24]

It wasn't long, though, before the house sparrow relinquished its status as the most hated bird in America. Their cousins from Europe, dark and at times mysteriously spotted, would see to that.

5

Across the Sea in Cages

EUGENE SCHIEFFELIN'S STARLINGS WERE BY NO MEANS THE only birds brought to North America. A lucrative bird importation business had existed for years in New York, where eager buyers waited impatiently for the latest crop of fashionable singers to put in their living room cages. By the late 1800s there were several large importing houses specializing in songbirds from Europe, plus retailers around the country. During the busiest seasons, the bird importers would hire twenty to forty travelers who went back and forth across the Atlantic on steamships, mostly to England and Germany. Once the birds were paid for and aboard the ship, the real work began.

"One experienced man can take charge of five large crates, each one containing two hundred and ten cages of birds, or a little over a thousand in all. Sometimes during the rush season the care-tender has five hurricane deckers to watch, or fourteen hundred cages and birds to look after during the long hours of the days and nights," one account published in *Scientific American* in 1898 said. "Feeding and watering over a thousand birds, and cleaning out their cages every day, makes up a routine of work on shipboard that begins at four o'clock in the morning and does not end until the late afternoon."[1]

Once in New York, the birds were taken to a shop, where they were nursed back to health if the trip had been onerous and were often trained to sing with the accompaniment of a reed organ or flute. Until the 1870s or so most of these birds, especially canaries, bullfinches, nightingales, and linnets, were destined for domestic cages. "But now the bird importers have a new demand for their stock," another 1898 article in *Scientific American* reported. "From all parts of the country bird societies and private individuals are purchasing the European song birds for the purpose of restocking the woods, fields, and parks of the country."[2]

Indeed, the acclimatization craze had taken hold far beyond the East Coast. In the spring of 1889, a few weeks after Schieffelin liberated his batch of seventy-two starlings at Central Park, a ship carrying hundreds of caged birds from Clausthal, Germany, steamed north up the West Coast, then headed east on the Columbia River at Astoria, Oregon, and continued to Portland. The freight in the holds, packed about twenty-two days earlier, included ten pairs of nightingales, twenty-three pairs of thrushes, twenty-two pairs of skylarks, four pairs of singing quail, and twenty pairs of starlings. Also aboard were dozens of bullfinches, robins, wrens, and finches. The noise below decks was cacophonous.

Portland was an upstart port city still rough around the edges, with rats, a dogged stench, and plenty of stumps and muddy streets to be navigated. The same year the birds arrived, the local newspaper, noting the unsanitary conditions of its sewers and gutters, called it "the most filthy city in the Northern States." The city was home to around forty-six thousand people but on the verge of a staggering population explosion, driven especially by eager immigrants looking for a fresh start in America.[3] They included a large contingent of Germans, some of whom clearly arrived with an enduring fondness for the birds back home. Amid so much tumult, should they be denied the sentimental trills and shrill calls of skylarks, finches, starlings, and thrushes?

The birds had been paid for by a Portland group calling itself the Society for the Introduction of Useful Singing Birds into Oregon. Sometimes it went by other names, including the Society for the Introduction of German Singing Birds into Oregon, the Portland Song Bird Club, and the Oregon Song Bird Society. At the heart was a group of German businessmen whose ringleader and secretary that spring was forty-four-year-old Christian F. Pfluger, a German-born shop owner and real estate broker who had an evangelical love of songbirds, preaching his gospel everywhere he could. Stocking the state with "these lovely songbirds," he said, should be considered a "publicly spirited enterprise."[4]

For the avian imports that arrived in Portland in May 1889, Pfluger and his group spent nearly $1,500 for 275 pairs of birds (more than $44,000 in today's dollars). "This large company of singers left Europe in perfect condition but they had a rough ocean voyage, and a number of them died . . . and many others were sick and worn out," one report said.[5]

From the ship, the exotic birds were transferred into large wooden cages and put on display to the public. The songbird society charged people to have a look and collected about $500. After four days the birds were released into the wilds of Portland and surrounding counties.

The efforts didn't go unnoticed by those outside Pfluger's group, especially by those who had also traveled far and missed some of the pleasures of home. "When the Easterner has spent enough time in Oregon to have become accustomed to the mild climate, luscious fruits, attractive scenery and numerous other good points, his attention finally is attracted to the scarcity of song birds as compared to the middle and eastern states," one local newspaper said. "There, on a spring morning, the air is filled with the melody of many a sweet songster's early carol, but in Oregon comparatively little music of this sort is heard. This fact has been so much regretted by bird lovers generally that action has at last been taken toward effecting a change in conditions in Oregon."[6]

Eighteen months after the release of the birds, Pfluger reported to the *Seattle Post-Intelligencer* that the skylarks and finches had "prospered and that the scheme has been a grand success." The city park hosted some of the newly imported goldfinches, and so did the county's poor farm just outside of town. Linnets were letting loose their "sweet songs" all over town. Nightingales had been spotted nesting forty miles south of Portland, chaffinches were near the ocean in southwest Washington, and "skylarks were seen in different parts of the state ascending into the air singing." Starlings, too, had been spotted breeding on a little island in the middle of the Willamette River three miles south of Portland.[7] The celebratory newspaper story, in the form of a letter from Pfluger, carried the headline "Music from Europe: Nightingales Lull the People of Portland to Sleep."

It had been such a rousing success that Pfluger and the other members of his society placed an order for hundreds more birds, including thirty-five pairs of starlings at $5.50 each, which were set loose in Portland in 1892. "The results of these two experiments were watched anxiously by bird lovers all over the United States. If they should prove adaptable to their new home, it would be the beginning of a great movement for importing European songsters," *Scientific American* reported in the second 1898 article, noting that while some of them, like the nightingale, ultimately didn't fare well, others seemed to thrive, including woodlarks, finches, and skylarks. "The most remarkable thing about these little strangers was their migration," the article continued. When the weather cooled in the fall, the birds flew south, instinctively following the warmer weather, some recorded in Southern California and Mexico. And when conditions warmed, many of the birds moved back north, settling in parts of Oregon and Northern California.[8]

It looked as though European bird songs were here to stay in some of the woods and fields of the Pacific Northwest. But alas, it didn't last. One by one, the hundreds of birds released by Pfluger dwindled and then vanished, unable to find a way to adapt to their new environment. Even the starlings, hardy and always ready to scrap for survival, couldn't make it. For a time, they were reliably seen in Portland and surrounding areas, including "nesting in the gilded ornaments" of the tower at the famed downtown Hotel Perkins. It was briefly hoped that the European starlings in the city might chase away the irksome crested mynas, another kind of starling that had also been imported to the region. But by the early 1900s the starlings introduced by Pfluger and his clubs were nowhere to be found. They'd disappeared, and no one knew why.[9]

Starlings wouldn't return to Oregon for another forty years or so. Meanwhile, in Ohio, another movement was afoot to get them on the ground and in the air.

ANDREW ERKENBRECHER, ALONG WITH HIS FATHER, MOTHER, and sister, stepped off the ship in New York City in July 1836 in search of a fresh start in the United States. His father, Heinrich, had been a cloth weaver back home in Germany, and fifteen-year-old Andrew was smart, energetic, and a fast learner. Four months later they were in Cincinnati and ready to plant new roots. Andrew learned English and soon began working, first in a sweet shop and then as a spice salesman. He saved money and bought a grain mill when he was twenty-two, later expanding to manufacturing laundry starch. Business was good, so he added another factory to his operation. He lost it all in a fire in 1860 and rebuilt the business a few years later on the banks of the canal near the village of St. Bernard, a suburb of Cincinnati. He again amassed considerable wealth, while gaining prestige and a sense of duty to his community. He even became the first president of the local telephone company. "His liberal hand was always open and remained open when it concerned the help of the oppressed or the accomplishment of works for the general public welfare," one remembrance said years later.[10]

In the spring of 1869 a call for help went out from the city. Like so many cities at the time, Cincinnati was beleaguered by thousands, perhaps millions, of caterpillars and worms. Its trees were beset by an army of these creatures that never seemed to tire of eating. "It has been suggested that the cheapest and surest way for Cincinnati to protect her shade trees from the plague of the caterpillar will be to adopt the plan pursued in New York and other Eastern cities—i.e., procure, and provide for, a large lot of English sparrows," the *Cincinnati Gazette* reported on April 28, 1869.[11] But who locally would do such a thing?

Andrew Erkenbrecher had always loved animals and especially birds. As a young man, he even kept caged songbirds. So it was no surprise that the answer came in the form of an announcement that a new society was forming in Cincinnati, one dedicated to "acclimatizing" foreign birds. But it wasn't just about acquiring insect eaters. The object, in official papers from 1873, was to introduce "all useful, insect-eating European birds, as well as the best singers." Secondarily, in response to violence against animals, the group vowed to protect domestic and imported birds "against the attacks of heartless men, and thoughtless boys." The society was open to "all persons bearing a good reputation."[12]

The intense Erkenbrecher, with a thick dark beard and piercing eyes, tended not to go halfway on anything. He was installed as president of the Cincinnati Acclimatization Society, sometimes called the Society for the Acclimatization of Birds, and threw himself into the work. In short order, the group pulled together $5,000 in cash and sent one of the society's board members, a man named Armin Tenner, to Europe to buy insect-eating birds and birds that could sing. While he waited for the birds to arrive, Erkenbrecher worked to secure laws to protect birds from being harassed, something he'd seen too often back home in Germany, especially at the hands of young boys.

Tenner didn't disappoint. He returned from Europe with about a thousand birds, including larks, orioles, robins, starlings, and of course, sparrows. For a while all the caged birds were housed and cared for in an abandoned mansion in a forested area called Burnet Woods. On a sunny spring day in May 1872 Erkenbrecher and a few associates went to the old house and set them all free. "One window was opened and a pair of nightingales appeared at the window, rested for a moment and then flew to the limb of a tree and, elevating their heads in the sunshine, burst out in joyous song," one remembrance said. "Within the next few seconds the birds fairly poured out upon the trees and shrubbery, filling the old woods with melody. It was one of the proudest days of Mr. Erkenbrecher's life."[13]

It was, Erkenbrecher later rejoiced, "a cloud of beautiful plumage" that led to "a melody of thanksgiving never heard before and probably never heard since." (John James Audubon was in the Ohio Valley a few decades before and was struck by the great variety of birds, including native songsters like wood thrushes, warblers, chats, and orioles, but perhaps their songs didn't fill the sentimental hole Erkenbrecher and others were trying to fill.) Erkenbrecher and Tenner evangelized about the benefits of the new arrivals, claiming they would not only protect orchards and shade trees but also enrich the lives of people hearing their tunes waft through the local woods and meadows, "which, in comparison with European countries, are so bare of feathered songsters."[14]

They were so enthusiastic about the results that over the next two years, they spent another $4,000 to import and release another three thousand songbirds, a similar mix of sparrows, thrushes, larks, finches, and starlings. It's a safe bet that the acclimatization society in Cincinnati brought more birds to America than any of their counterparts elsewhere in the country.

While the effort was celebrated in some corners, not everyone was pleased. The Cincinnati Society of Natural History, full of men married to a belief in science and the protection of the natural world, pointed out the wrongheadedness of releasing foreign birds with such little concern for unintended consequences. They were particularly worried about native birds like martins, wrens, and swallows and noted that if the new birds were able to gain a foothold in their new homes, they could "only do so at the expense of the native ones." The concern was slightly different in the city, the society said, where native birds were relatively few: "Here the sparrow lends an attractive air to the monotony of brick walls and cornices, even if he does disfigure the latter with his bulky nests, and scatter his lateritious cards somewhat too numerously over our window sills and doorsteps." Even if there were a complex mix of costs and benefits, the Society of Natural History said the bottom line was simple: "Take care of what birds we have, by a judicious preservation of thickets and other abiding places, and nature will provide effectually against the calamity of an ornithological vacuum." In other words, stop screwing around with Mother Nature, and she will respond with what was needed.[15]

So perhaps there was some relief among the society that nearly all the birds released by Erkenbrecher and Tenner, despite abundant funding and enthusiasm, were unable to survive in Cincinnati. Erkenbrecher eventually founded the venerable Cincinnati Zoo—it's easier to ensure the long-term survival of species in cages rather than in the wild—but the sound of his beloved birds from back home went very quiet. The exception? Not the starlings but the sparrows, which thrived, spread, and delivered on their promise to be a nuisance for generations to come.

Beyond Ohio, others were busy, too, manifesting the lofty ideals of acclimatization. Around the same time the birds were being released in Cincinnati, the Society for the Acclimatization of Foreign Birds in Cambridge, Massachusetts, released European goldfinches at Mount Auburn Cemetery, a few miles west of Boston. The birds and their offspring were seen for decades in New England and beyond. In 1891 the Country Club of San Francisco—which formed mainly to introduce brown trout from Europe into the streams of California—set free sixty-seven pairs of mockingbirds brought from Louisiana, along with several species of foreign songbirds like European goldfinches and nightingales.

BACK IN NEW YORK CITY, EUGENE SCHIEFFELIN EAGERLY READ reports about bird releases elsewhere in the country, with a particular interest in what was happening in Oregon. In the early 1890s he struck up a correspondence with Christian Pfluger, the driving force behind the songbird project in Portland. Schieffelin said he was "intensely pleased" with those efforts. "Were it not for the mountains between us," he wrote in neat cursive, "I might hope that the birds you set free might eventually spread throughout the whole country but I fear the Rocky Mountains will be an obstacle to the realization of such hopes."[16]

Schieffelin detailed to Pfluger his long history with introductions in the East and said that his American Acclimatization Society was now defunct, leaving him to pursue the hobby on his own. He'd recently set loose European species of thrushes, goldfinches, and chaffinches with mixed success, he said. Starlings were another matter. "I am happy to say that the starlings have bred about New York this summer in increased numbers and are undoubtedly established," he wrote in August 1894.[17]

Any criticism cast his way about the perils of introducing foreign birds on American soil fell on deaf ears. Schieffelin instead offered Pfluger a grand vision for importing even more birds: "In our country with all its variety of climate and conditions, the feathered tribes of almost the entire world could be introduced with a little tact in releasing them at the proper time and spot."[18]

6

Lessons from Down Under

HAD EUGENE SCHIEFFELIN, CHRISTIAN PFLUGER, OR ANDREW Erkenbrecher bothered to wonder about the wisdom of importing and releasing wildlife from other lands, they would have been well served by a trip Down Under in the last half of the nineteenth century.

Perhaps most famously, European rabbits were set loose in Australia in the 1800s to give hunters something to shoot. Apparently, dingoes and kangaroos weren't cutting it. Although some rabbits were already on the continent, the population in the wild exploded after the midcentury introductions. Thousands quickly became millions. Crops and pastures succumbed to the "gray blanket," as the rabbit masses were sometimes called. By the end of the century rabbits occupied more land than Texas, California, and Montana combined—and millions had been spent on bounties, poisons, and fences to keep them out. "Nothing has yet been found that will effectually exterminate the pest," one disheartened chronicler wrote.[1]

Losses in the colony of Victoria alone were estimated at $15 million, and frustrated businessmen facing ruin soon pivoted to selling rabbit pelts. More than twenty-nine million were exported from the colony for hats and other items over the course of a decade. Relief remained elusive. "The colonists are at their wits' end how to repair the evil," a report said in the summer of 1889.[2]

Around the same time, rabbits were also introduced into Tasmania and New Zealand, where they "spread like a scourge."[3] Predictably, someone had the idea to import yet more foreign animals to deal with the pesky imported rabbits. In New Zealand, thousands of weasels, ferrets, stoats, and cats were purchased and released in hopes of controlling the rabbits. The predators put a dent in the rabbit population but were soon feasting on the native birds too. Instead of being a solution, it had only compounded the problem. It was, one New Zealand farmer and conservationist sniffed, an "attempt to correct a blunder by a crime."[4]

Undeterred by the fiasco of the rabbits, colonists soon turned their attention to bringing in birds from faraway places. Sentimental colonists who were confused and put off by the strange creatures in their new homeland found themselves yearning for the natural comforts of the Old World. "At night no more he hears the delightful warblings of the queen of the songsters—the charming nightingale," the homesick British Army assistant surgeon Thomas Bartlett wrote in 1843 in Australia. "The hoarse croaking of the offensive bull-frog, and the incessant buzzing of the hideous mosquito, he takes in exchange for the gladdening tones of England's fairy songsters."[5]

One of the remedies seemed obvious. Records show that some of the first starlings were released in Sydney in 1856, followed by more in Melbourne in 1861, Brisbane in 1869, South Australia in 1879, and still more near Sydney in 1880. Not to be outdone, New Zealand went on its own starling spree, setting them loose in 1863, 1867, 1872, 1874, 1876, 1877, and 1888. And those are just the releases that were documented.

The starlings slipped easily into their new homes, thriving and breeding and spreading constantly into new territories, ever on the hunt for food. Regret inevitably followed among the citizenry, as cherry, apple, and wheat crops were decimated, and local birds struggled against their prolific and highly adaptive new neighbor. The land of kangaroos, koalas, wallabies, and dingoes suddenly felt under siege by a rambunctious and sparkly black bird small enough to hold in your hand. "Had I been asked fifteen or twenty years ago what I had to say, I would probably have recommended their introduction," an Australian government biologist, R. Helms, said in 1898 about starlings. "But not so now. My experience has taught me better."[6]

In some cities and suburbs, starlings became ubiquitous, tumbling down chimneys or cramming onto roofs, where they nested among acquired sticks and grasses. Often they brought little mites with them, which soon infested homes, beds, and clothes. The tiny bugs bit people, sucked blood, and left behind red, raw rashes. "The only way to cleanse the house thus infected is to get rid of all the rubbish in the roof, close up all openings through which the birds can enter the eaves, and thoroughly disinfect the whole place," one report from Australia said. The starlings also ate their way through apple and citrus orchards, leaving some conflicted about their feelings for the birds because they ate insects as well. "It is not of much advantage to the orchardist if the starlings destroy his fruit crops in the summer, for as a general thing the insect pests would have left him something," the report noted.[7]

Livestock wasn't immune either. At first ranchers were pleased to see starlings on the backs of their sheep, picking off ticks and other insects. Unfortunately, the starlings rarely did anything in moderation. "In flocks of sheep, they alight on the sheep's back, and hurt the evenness of the wool to a great extent," said the same report. "It has been noticed that so terrified are the sheep when a flight of starlings approach that they run away as though rounded up by dogs."[8]

In 1893 the government in the colony of Western Australia, worried by what was happening elsewhere on the continent, passed a law banning the importation of destructive animals and birds. Initially, it included sparrows and rabbits. In 1896 starlings were added to the list. The United States wasn't far behind.

CLINTON HART MERRIAM WAS BARELY A YEAR INTO HIS NEW job as the chief of the Division of Ornithology and Mammalogy (later called the U.S. Biological Survey) in the U.S. Department of Agriculture when he voiced his concerns about importing and releasing nonnative wildlife in the United States. In 1886 Merriam, a zoologist, ornithologist, and doctor who had just left his medical practice for this new government position, was no doubt aware of what had happened with the explosion of sparrows in the United States, the problems caused by rabbits and starlings in Australia, and the terror that followed the release of mongooses in Jamaica. "The question of the desirability of importing species of known beneficial qualities in other lands is one which sooner or later must force itself upon our notice," Merriam said, adding that particular attention should be paid to songbirds.[9] He suggested that all future introductions be approved or rejected by the U.S. Department of Agriculture.

His plea fell largely on deaf ears for more than a decade, until California-born naturalist Theodore S. Palmer began sounding the alarm. Palmer had led a federal expedition to explore the biology of Death Valley in 1891 and eventually spent much of his career at the U.S. Biological Survey and as an ornithologist fighting to stop the slaughter of native birds in America. In 1898, when he was the assistant chief of the Biological Survey, Palmer issued a damning report called "The Danger of Introducing Noxious Animals and Birds." In it, he roundly condemned the acclimatization societies, including those in the United States, while enumerating the damage worldwide from the importation of bats, rabbits, sparrows ("one of the worst feathered pests"), starlings, and others. He included pictures, maps, and detailed descriptions of what he termed "the evils of indiscriminate and ill-advised acclimatization."[10] He chastised the short-sighted view of those deliberately releasing foreign species, noting that not only would the human world be inescapably altered but the natural world would be too. He took particular aim at those importing and releasing birds. "In the eagerness to acquire new birds, the risk of importing undesirable species has been overlooked, and even the lesson of the English sparrow has not been enough to impress on the general public the dangers of ill-advised acclimatization," Palmer wrote in a separate report the following year.[11]

Any chance of avoiding the upheavals in Australia and New Zealand required the United States to take a firm stance by giving complete control of wildlife imports over to the Department of Agriculture. Palmer didn't have to wait long. In the spring of 1900 Congress took up the Lacey Act, a sweeping law named after Iowa congressman John Lacey that was meant to protect native wildlife, rein in rampant poaching and trafficking, and reduce the movement and importation of foreign species. It specifically banned the introduction of European starlings, sparrows, mongooses, fruit bats, and other animals.

The bill drew support from farmers, horticulturists, sportsmen, and bird lovers. Opponents included those who stood to lose out because of tighter restrictions, especially those in the business of making women's hats splashed with bird feathers. But the need was obvious, Lacey argued. One had only to look at what had happened with the release of mongooses in Jamaica, which proved more of a nuisance than the snakes that they were imported to wipe out, Lacey said. Other examples included the rabbits set loose in Australia and Russian thistle, which had taken root in North America and become infamous as "tumblin' tumbleweed" around the West. By the time the Lacey Act was being debated in the first half of 1900, most of the nightmarish stories about imported birds in the United States were dominated by house sparrows, as they had a few decades' head start on starlings, which were only beginning to spread around the East.

"If this law had been in force at the time the mistake was made in the introduction of the English sparrow, we should have been spared from the pestilential existence of that 'rat of the air,' that vermin of the atmosphere," Lacey said during one of the final debates over the bill. "But some gentlemen who thought they knew better than anybody else what the country needed saw fit to import these little pests, and they have done much toward driving the native wild bird life out of the States."[12]

The Lacey Act passed and was quickly signed by President William McKinley. It became one of the most important wildlife laws in the first half of the twentieth century, but for starlings, it came years too late. The genie was already out of the bottle.

7

Occupation

HISTORY DIDN'T RECORD WHAT EUGENE SCHIEFFELIN THOUGHT about passage of the Lacey Act in the spring of 1900. Did he take it personally that the new law specifically outlawed the release of sparrows and starlings, two birds he'd spent a good deal of time and money importing? Was it a strange point of pride to have his work singled out in America's first wildlife protection law? Did he shrug it off as the work of shortsighted politicians who did not have an inkling of the beauty of the starlings' flights and colorful songs?

What is clear is that the starlings were unfazed by the Lacey Act or any other effort to keep them in check. *Sturnus vulgaris* was already firmly in place and had no intention of going away.

By 1900 Schieffelin's starlings were in Flushing, New York, and East Orange, New Jersey. The following year they were spotted near Odessa, Delaware. In 1904 they reached Bucks County, Pennsylvania, and Rye, New York, and in 1905 they showed up in Philadelphia.

In the fall of 1907 a story by ornithologist Frank Chapman was published in the magazine *Country Life in America*. His piece, called "Our New Bird Citizen," was all about the starlings' arrival in the United States. He tried to calm emerging fears about what this new, unfamiliar resident might do. "The starling is not so prolific, hardy or adaptive as that 'rat of the air,' the sparrow," Chapman wrote. In fact, he said, starlings could even liven up some places. "Particularly in winter, when the sparrows seem to be our only bird neighbors in many suburbs, where the starling is there is added [a] bit of life in the landscape and a note of joy in the bleakness."[1]

In the winter of 1905–6 a strange bird was seen in New Haven, Connecticut, nearly in the shadow of Yale University. Although from a distance it looked like an ordinary blackbird, a closer look revealed dark plumage with a green gloss and, when the sun hit just right, spangles of brown. "But most striking and characteristic of all was its slender, bright-yellow bill," according to an account from Yale's superintendent. The bird not only occupied the trees but walked the streets with confidence, "carrying itself with an air of aloofness and importance." The school official struggled to identify the visitor until the state ornithologist, Herbert K. Job, let him know that the European starling had come to the City of Elms.[2]

The small colony in New Haven became hundreds in a few years. And in the nearby marshes of West Haven, a few "small parties" in 1909 morphed into two thousand a year later. Job spent several years tracking the starlings in and around New Haven and sounded a note of alarm about the early incursions. "The wave of starlings has steadily gathered headway and will no doubt in due time inundate the whole United States," Job wrote in a detailed report titled "Danger from the Starling," published in *Outing* magazine in 1910, which included the account from the Yale superintendent. "But have we another English sparrow pest on our hands?" he asked. "People are everywhere beginning to inquire."

The Connecticut ornithologist provided one of the first in-depth accounts of starlings occupying a new region of North America. Job, trained as a Congregational pastor, favored cameras over guns when confronting birds, viewing them as creatures of God that deserved protection rather than elimination. Even so, he didn't bother to hide his disdain for the starling, calling it "a most filthy bird" and one apt to become a great nuisance. He detailed its migration habits, food preferences, calls, and skittish disposition at the first sign of trouble. Most of all, he tried to convey simply how prevalent starlings had become for everyday people in town.

"As the weather becomes cool they are often seen walking in the streets or in the backyards looking for garbage. They mingle with the poultry, and, after taking their toll of grain or scraps, perch on the shed roof, dress their glossy plumage, and whistle and warble," he said.

In the winter afternoons, just a few hours before sunset, Job watched as the starlings began to congregate in small groups in the tall elms near a church. "The company grows till the trees are fairly black with them. A starling concert is in order, attracting the attention of people passing by," Job wrote. As darkness fell, the birds made their way into the church tower. "I have been up there at night with a lantern. Every available niche or perch on the slats, joints, or timbers is crowded with them. The light attracts them, and clouds of them fly into my face, fairly blinding me. When I need specimens to study, I can collect all I need in short order. They are active, muscular birds, and put up a great fight to escape, when laid hold upon."

The starlings proved to be prolific breeders, capable of nesting in towers, barns, sheds, bridges, and under eaves—essentially anywhere there was a deep, dark hole where they could lay their delicate, pale blue eggs. But that often came with a steep cost to native birds that lost their nesting holes to the aggressive newcomers, Job said. He'd seen starlings bully their way into the homes of larger, stronger birds like flickers, taking advantage of the work that had been done before they arrived. In some cases, the flicker is "degenerating into the carpenter-slave of the starling," he said. Job described one instance where starlings drove a pair of flickers out of their hole, and a local resident had to shoot nineteen starlings before the flickers would return home. He'd even seen that dynamic on the street where he lived, witnessing the plight of flickers and woodpeckers robbed of their homes. "Finally, as though in despair," Job wrote, "they came and built right by my front door, as though making demand upon the State Ornithologist to protect them in their rights."

He kept a close eye on reports that the starlings were spreading across the East and couldn't help but wonder what future they portended. It's possible, he speculated, that starlings could be tolerable in moderate numbers, especially if they ate up insects, grubs, and larvae in the spring and summer. But swarms in cities and neighborhoods would be a mess, and farmers might have a darker view of them once starlings turned from insects to crops. And he couldn't stop worrying about what might become of the local birds he loved so much. Job lamented that by this time there were so many starlings, and they were spreading so fast in so many places, that "it would now be impossible to exterminate them."[3]

The march continued into Springfield, Massachusetts, in 1908 and then Providence, Rhode Island, in 1910, where starlings roosted in the roof peaks of cottages and behind the lattice work of homes. By 1915 they were seen throughout much of southern New England and in parts of Maine, New Hampshire, Vermont, and Virginia.

Around this time, Massachusetts ornithologist Edward Howe Forbush was in touch with more than a hundred people in five states who were tracking the starlings' progress. "All assert as a result of their observation that it is increasing, and most of them say that its accession is rapid," Forbush said. For example, people reported seeing one thousand to three thousand starlings in one area in the fall of 1909 and, in the same place the following year, eight thousand to ten thousand. Typically, starlings were outnumbered only by sparrows and robins, he said. But that wouldn't last long.

During the spring breeding season, the starlings tended to be fairly quiet and secretive, but in the fall, the flocks were hard to miss as they swarmed fields in search of food and shimmered in the sky in dark clouds of wings and feathers. Forbush, an intrepid naturalist who had long observed the starlings, began constructing a portrait of these new residents and found much to admire, noting they were as hardy as a crow and as wily as any bird when it came to avoiding the danger of predators. "The starling's physical fitness for the struggle for supremacy is seen at once on an examination of its anatomy," the ornithologist wrote. Not only was the bird exceedingly tough and wiry, but its bill—"its principal weapon of offense and defense"—seemed particularly useful. It was straight, heavy, and strong enough to dig for food in the hard ground, "nearly as keen as a meat axe, while the skull that backs it is almost as strong as that of a woodpecker." Forbush also observed, "Mentally the starling is superior to the sparrow and active in the face of any foe that it can master, it shows the acme of a caution and intelligence in its relations with man or any other creature too powerful to master."

Even though starlings were still largely confined to the Northeast, Forbush understood that this species was quickly spinning out of control, and any attempts at this point to kill it off—perhaps by poisoning or other means—would likely mean large-scale collateral damage among native birds, like meadowlarks. Still, it wasn't hard to see how the starlings might upset the natural way of things, both for people and for native wildlife, especially other birds, which starlings sometimes attacked and bullied. "The starling is a sphinx-like bird and ordinarily treats other birds with a sort of contemptuous tolerance," he wrote.

Starlings liked to nest in cavities and were quick to adapt to their human environment. It wasn't long after Schieffelin's introductions that they were spotted under eaves and in church steeples, dovecotes, electric light hoods, and bird houses. And when those weren't available, the starlings invaded existing tree cavities occupied by other birds. Like his fellow ornithologist Job in Connecticut, Forbush was already seeing signs that the pugnacious starlings would compete for nesting places with native birds, squatting behavior sure to drive away flickers, woodpeckers, crested flycatchers, martins, bluebirds, swallows, and wrens. The starlings' strength helped, but so did their ingenuity and sheer numbers. Forbush described one scene he witnessed:

> As the starling comes, native birds, whose nesting places it covets, must go, and many of these birds are more desirable than the starling. The skillful manner in which it evicts the flicker inspires the observer with a certain admiration for its superior strategy and prowess. The starlings quietly watch and never interfere while the flicker digs and shapes its nesting place in some decaying tree; but when the nest is finished to the satisfaction of the starlings it is occupied by them the moment the flicker's back is turned. On the return of the flicker a fight ensues, which usually results in the eviction of the particular starling then in the hole, which, however, keeps up the fight outside while another enters the hole to defend it against the flicker, which, having temporarily vanquished the first, returns only to find a second enjoying the advantages of possession.

Forbush continued: "The moment the flicker gives up vanquished, the starlings molest it no more, allowing it to hew out another hole, either in the same tree or in one nearby, when a similar fight ensues with more starlings; and so the flicker is driven literally from pillar to post, until it has prepared sufficient homes for the starlings in its neighborhood and all are satisfied, or until it gives up in disgust and leaves the vicinity of its aggressive neighbors."

When it came to the starling's effects on people, most of the problems revolved around its appetite for food crops. Forbush collected dozens of reports of starlings that had devoured vast crops of cherries, pears, apples, strawberries, lettuce, and radishes. "Sometimes they strip a tree completely and then go to another," Forbush wrote about starlings' appetite for cherries. "In other cases they feed in a desultory way, taking toll from all the trees in a neighborhood." But it wasn't all bad. While starlings put a dent in crops, they also feasted on insects. In 1910 a federal scientist examined the stomach contents of 102 starlings, including about 40 young ones, and found bits of caterpillars, spiders, millipedes, inchworms, weevils, grasshoppers, crickets, and beetles.

We can almost hear Forbush sighing as he tried to find a way to deal with the new reality of starlings in the United States. Even if they might provide some benefits, the situation was anything but simple. There was likely no sense in trying to eradicate them at this point, he said, and the best options might be a vigorous defense, including nets on top of crops, igniting Roman candles, and hiring boys to run around clapping boards together to drive them away. "Whether its presence will result in more good than harm will depend largely on the ratio of its increase," he said, striking the same sense of resignation as other ornithologists. "The starling is here to stay, and we must make the best of it."[4]

IN THE SUMMER OF 1911 THOUSANDS OF STARLINGS, SPARrows, and other birds took up residence in a tony, tree-lined enclave of Montclair, New Jersey, known as the Crescent. Locals complained they were noisy, smelly, and left a huge mess. One night in mid-August a garbageman named Peter Stevens showed up in the neighborhood with a double-barreled shotgun. The *New York Times*, in a front-page story the next day, offered a description of what happened next: "Soon after the first shot many of the wealthy residents poured out of their homes to aid the man in the work, supplying lanterns and baskets to pick up the dead birds." Hundreds of birds were killed in the course of a few hours, the newspaper said. Others were wounded or died later, falling from the trees and onto the road and lawns.[5]

"The residents of The Crescent, who had complained to the Shade Tree Commission and the police authorities because the birds had disturbed their slumbers at night, aided in the work of destruction by killing the wounded birds that rained from the trees," came a follow-up story in the *Times*. "The people of the town are divided over the execution of the birds. Even the sparrows have champions."[6]

The story of the bird killing generated quick outrage, along with the arrest of Stevens. The garbageman said he'd been offered $2 by the town's chief forester to kill starlings in the neighborhood. Because killing starlings was illegal at the time in New Jersey, Stevens was fined $100 by the local justice of the peace. He couldn't afford to pay, so he was sentenced to ninety days in jail. Most city officials apparently didn't make much effort to free Stevens, who was Black, so he sat in the local jail. The town's attorney, though, caught wind of the case and went looking for the state's supreme court justice to free the man, first taking a train to Trenton and ultimately tracking him down on a golf course. Stevens was soon out on bail.[7]

Members of the local Audubon Society investigated the incident and found that in reality, only twenty-eight birds had been killed, a mix of starlings and sparrows. Still, the publicity around the "war on starlings" in Montclair, as some soon dubbed it, made the papers across the country and generated a wave of indignation. "A cry of protest rose from all parts of the United States in such a volume that Montclair was dazed and the residents of the Crescent determined that come what might, they would never venture upon such means again."[8]

The starlings left in the fall but returned to Montclair with a vengeance the next summer, with some thirty thousand congregating in the stately maple trees alongside grackles and robins, "making life miserable" for local residents of the Crescent. The birds' droppings defaced the sidewalks and houses, one resident complained, "with no compensation in melody." In fact, the noise was "about as musical as the hum of a sawmill or the din of a boiler factory." The neighbor added that any bird lovers, presumably those opposed to lethal action against the starlings, were welcome to see the situation for themselves, "but I would counsel them to wear their oldest raiment, hoist umbrellas, and fasten clothespins on their noses."[9]

At first the city tried trimming some of the trees along the avenue where the birds were roosting. When that didn't work, people tied sleighbells to the branches, hoping that would keep the birds away. In defeat, the city called out the guns.

In early August seven deputies with the state Game and Fish Commission walked through town, firing shotguns into the trees. Given the controversy over the dead starlings a year before, their guns were loaded with blanks, the noisy powder meant to merely scare the birds away. The ruckus went on for hours as deputies fired their guns and the birds took off, only to land in a clump of trees nearby. Back and forth they went as a crowd gathered and dismay grew. The birds weren't really going anywhere.[10]

The following spring New Jersey legislators, responding to the situation in Montclair, passed a law allowing starlings to be killed in hopes of giving locals some relief from the chirping and chattering keeping them up at night. The shooting began later that year, driving the starlings away but never for good.

THE INCIDENT IN MONTCLAIR PRESAGED DISCUSSIONS THAT were about to take place in much of the East and later across the country: Should starlings be killed or left alone? The presence of starlings never failed to get people talking and pursuing their own methods to keep the new birds in line.

In Pennsylvania, state game wardens were told to keep a sharp eye out for starlings and kill them on sight to keep them from gaining a foothold. A couple of months later hundreds of starlings descended on Stamford, Connecticut, including near the home of a man named Pasquale Palo. He grabbed his gun and fired at the flock, killing one starling and injuring another. He was fined $14.69 and found guilty of two counts: killing a starling and having a starling in his possession. The case made the front page of the *Bridgeport Evening Farmer* later that day.[11]

In Hartford, Connecticut, there were wild reports of a city under assault by millions of starlings that left citizens sleepless and of unwanted squatters occupying the highest points of St. Joseph's Catholic Cathedral. The problem was turned over to the city forester and the local Bird Study Club. Their solution: "Teddy bears are to be fastened to the tree where the starlings have nested. Tonight a few rockets were fired through the trees, and if there is any noticeable result a fusillade will be discharged each evening for a week or more." A dozen men were dispatched to one street, where they fanned out down the block and, at a signal, set off a volley of heavy-duty Roman candles directly into the trees. In hopes of improving results, the next round included a few brave men who climbed high into the trees to set off the explosives. It was a spectacle. Meanwhile, hundreds more were seen in and around Springfield, Massachusetts, "frequenting the spires and cupolas of the churches and public buildings."[12]

In Bridgeport, Connecticut, in March 1914 a flock of a hundred or so starlings started showing up every day. "Their cheery calls have been enjoyed by the bird lovers in the neighborhood. Last Sunday much shooting was heard in this section and it was reported that the hunters were killing starlings," a farming newspaper reported, noting that while some objected to the birds, their appetite for potato bugs "should be recommendation enough to protect them from all harm."[13]

There were reports of starlings in Hadley, Massachusetts, and Westchester, Pennsylvania, and then in New Hampshire, Vermont, southern Maine, and Norfolk, Virginia. The birds moved along the coastlines, up valleys, and along rivers, places where they could fly with an easy line of sight to avoid predators. By 1916 a few of the most intrepid explorers flew up and over the Allegheny Mountains, and the westward expansion accelerated. In November 1917 a man near Savannah, Georgia, killed a starling, one of the first documented south of Philadelphia.[14]

Rochester, New York, had them and so did Wheeling, West Virginia; Ipswich, Massachusetts; and parts of Ohio. Sometimes they traveled with red-winged blackbirds and cowbirds in a companionable partnership among species of similar sizes and dispositions. "A Hopkinton [Rhode Island] man, Simon Kenyon, killed ten starlings with one shot the other day," came a report on New Year's Day in 1920.[15]

In Yonkers, New York, a car dealer lost his temper at seeing a group of starlings attacking some local songbirds near his business. He grabbed his rifle and started shooting at the flock, killing one before he was arrested.[16]

Eugene Schieffelin wasn't around to see any of it. He'd left New York City in the early summer of 1906 to spend some time in the ocean air at Newport, Rhode Island. He had a stroke and died three weeks later, at the age of seventy-nine. The *New York Times* ran a short article about his death, noting that his father had been an important lawyer during the previous century and that Eugene had been a member of several prominent clubs and societies.[17]

There was no mention of his starlings.

8

European Origins

ON MAY 17, 1903, CLINTON G. ABBOTT WANDERED INTO NEW York City's Central Park to look at birds. He was about to graduate from Columbia University and move on to graduate studies at Cornell. Later he became a full-fledged and well-known naturalist, ornithologist, and director of the San Diego Natural History Museum. That day, though, he was bound for the fresh air of Central Park. Within twenty minutes, he was able to spot five species of birds imported from Europe: chaffinch, goldfinch, greenfinch, house sparrow, and starling. He'd seen some of them before, including a flock of more than a hundred starlings in the Columbia neighborhood. But he was struck by the array of foreign guests parked in the trees that spring day. "From an ornithological standpoint we must surely speak of the European invasion of America," he wrote in an article titled "European Birds in America."

At the time, ornithologists, bird lovers, farmers, and others were still trying to determine whether the starling would be a friend or foe in America. Naturally, many looked to their native land across the Atlantic for clues. Growing up, Abbott had split his time between England and America and was well versed in the reputation and activities of starlings. "Its rapid increase offers only too evident proof that it will not be many years before it gains in this country the position which it holds in England, in being second only to its compatriot, the House Sparrow, in dominion over the land," Abbott said. He predicted hardships coming for America's bluebirds, martins, crested flycatchers, woodpeckers, and others that nested in holes.

"To be sure," Abbott continued, "he possesses a song, but it is third-rate at best, and the beauties of his plumage can be appreciated only as close quarters; let us only hope that he will not, under the new conditions, change his diet, which at present is chiefly insectivorous, or woe betide the farmer beneath the ravages of his vast winter flocks!"[1]

THERE'S A STORY THAT GOES LIKE THIS: ON A SUNNY SATURday in the fall of 1621, a war of starlings broke out in the Irish city of Cork. The birds had been amassing the previous week, with two massive bunches of "stares," as starlings were often known, roosting noisily in the trees and on buildings, one in the west and one in the east. Occasionally a group of twenty or thirty from one army would venture to the other, where "they would fly and hover in the Air over the Adverse Party, with strange Tunes and Noise, and so return back again to that Side, from which it seemed, they were sent."

The excursions into enemy territory continued until the morning of October 12. At around 9 a.m. a sudden noise rose up over Cork, as birds from both sides launched into a violent clash with one another. "Upon this sudden and fierce encounter, there fell down into the City, and into the Rivers, Multitudes of Starlings or Stares, some with Wings broken, some with Legs and Necks broken, some with Eyes picked out, some with their Bills thrust into the Breast and Sides of their Adversaries."

On and on it went throughout the day, a horrible war overhead with the carnage dropping from the sky around the stunned citizens of Cork. The fighting stopped a little before sundown. The next day, a Sunday, the starlings were mysteriously nowhere to be found. On Monday they were back and the battles resumed, with all manner of mangled and slaughtered starlings toppling onto houses and into the rivers that cut through the city. And then it was done. The war was over, but it was never known which group of birds won.[2]

The starlings in Cork that day, just like the ones Schieffelin released in Manhattan, were part of the Sturnidae family, a group of more than a hundred species of starlings and mynas around the world. It's a songbird family full of aggressive chatterers with bills that curve slightly downward, skinny strong legs, and pointed wings that make them expert fliers that can zig and zag at a split second's notice. These are birds that like to make noise and gather in spectacular numbers. They are roosting birds, benefiting from millions of years of evolution that provided perfect perching feet with three toes pointed forward and one to the back.

It's also a diverse family of birds that shows up around the world. Some are small, perhaps six inches or so, while others reach more than two feet. Hildebrandt's starling in Kenya and Tanzania, for example, has a striking look, with a blue head, orange breast, and glossy blue-green tail. The superb starling in Kenya has a black head and a white bar that separates its red-orange belly from its blue breast. The wattled starling is mostly gray, and the breeding male has a black forehead, a bit of yellow skin that shows through on the head of some, and a throat wattle. The chestnut-tailed starling in India and Asia comes in a mix of gray, orange, and black. The rosy starling stands out with its pink and black plumage.

By far the most numerous starling on the planet—certainly the most derided and discussed—is the common starling, or European starling, as it's often called in North America. Its formal name, *Sturnus vulgaris*, is simply Latin for "starling" and "common."

Native to parts of Europe, Asia, and North Africa, the common starling has spent most of its long existence confined to those areas, gathering in lowlands and developing a knack for eating a wide variety of food. Particularly handy is its unusual beak, which has strong muscles not just to chomp down but to pry open, allowing it to probe the ground and widen holes in search of insects or other foods. In Europe as well as in North America, starlings often move through valleys and take up residence near cities, towns, and other human settlements. Many of them migrate, but it's often haphazard. As they move into a new territory, a few pioneers typically venture forth first and then the larger population follows, usually setting up shop about five years later. For thousands of years, starlings and people have lived as neighbors, sometimes peaceably, sometimes not. Emotions have been just as apt to be fueled by irritation as reverence.

English historian John Aubrey, writing in the 1600s, speculated that the builders of Stonehenge were so smitten with starlings—he called them "Druids' birds"—that they deliberately left holes in the joints between upright stones and lintels for them to nest in. Indeed, the Welsh word for starling is *drudwen* or *drudwy*.[3]

In ancient Rome, massive flocks of dark-colored starlings often filled the sky with dazzling shapes and patterns. Augurs, or priests, would search these otherworldly bird formations to divine the will of the gods, which in turn informed decisions about wars, domestic issues, and even where Rome itself should be built. The practice was called "taking the auspices."[4]

In his 1909 book, *The Old Town*, Jacob A. Riis recounted his childhood in rural Denmark. While drifting down a river on summer evenings, he'd listen to the chatter of thousands of starlings roosting in the reeds. Along the way, he'd shout and clap his hands, rousting the startled birds from their rest in immense, thunderous clouds. It brought only delight. "We loved him and gave him sanctuary. And he helped the farmer in turn by ridding his field of pests. Where a flock of starlings settled down for luncheon, no wriggling thing remained to tell the tale," Riis said. "Sparrows are cheeky tramps—the starling a friend."[5]

Not everyone in Europe held starlings in such high regard. As in North America, the situation was complicated. Some farmers welcomed starlings for their appetite for insects on the grasslands but cursed them for tearing through fruit farms, feasting on cherries, apples, plums, currants, and raspberries. Farther south, in France, the birds lustily consumed olives. In Switzerland, it was reported that a black cloud of starlings robbed a vineyard of a year's worth of crops in fifteen minutes. As more land was converted to agriculture, human attitudes toward starlings soured further. "In my fruit field, I do not suffer very much from blackbirds and thrushes, nor do I grudge them their toll in return for their song," an English farmer named S. H. Goodwin said in 1908. "Only one bird is dangerous to my crops—that is the starling." As soon as the young hatched, they arrived from the marshes. It was a staggering sight, he said. "And they come in millions; in flocks that darken the sky. Their flight is like the roar of the sea, or like the trains going over the arches," Goodwin said, noting that starling numbers had been increasing over the past forty years to the point of being intolerable. "The starling is a terror, and life around here is hardly worth living; you must have a gun always in your hand, or woe betide the cherries; they come in thousands."[6]

Always, though, there were those who fell under the spell of starlings—people who were willing to overlook any damage they caused to human goods and instead saw them as a force of enigmatic wonder and brilliance.

EDMUND SELOUS WAS A BRITISH ORNITHOLOGIST AND WRITER at the turn of the twentieth century but didn't exactly fit the mold. He was cranky, funny, a committed Darwinist, relentlessly curious, reliant on his eyes and ears, and given to steering clear of other ornithologists or even reading their work, lest he be unduly influenced by ideas about birds that he hadn't observed during his time in the field. His early books were mostly a compendium of his scattered diary entries, the writing precise and informed by a seething fascination for all things wild and misjudged, including starlings. "Selous was a genius, lonely, misunderstood, unappreciated—a tragic figure in his old age with his passion for the lovely wild things of the earth that are being relentlessly driven to their doom by cruel and indifferent man," said one remembrance written in 1935, a year after his death.[7]

In 1899, when he was in his early forties, Selous moved to the tiny village of Icklingham, on the eastern side of England in Suffolk County. The land around him was flat, sandy, speckled with trees, and carved up by charming rivers and streams. Best of all, it was often ripe with birds. Selous made almost daily pilgrimages into the fields and woods near his country cottage, always with a keen eye and a ready notebook. Not that it was always fulfilling. "To be a field naturalist in England, is to be a field ornithologist, and more often than not—I speak from experience—a waster of one's time altogether," he wrote in his 1905 book, *Bird Life Glimpses*, adding, "Be assured that you will often feel immensely dissatisfied with the way in which you have spent your day."[8]

Still, he found plenty to see and report on in his three years in Icklingham. He filled pages and pages with detailed seasonal remarks on wood pigeons, herons, rooks, green woodpeckers, nightjars, thrushes, and peewits. He saved his best for dissecting the behavior of the starlings that populated the region, basking in the wild poetry of these mysterious birds, which often arrived in "numberless numbers" above the horizon: "A black portentous cloud shapes itself on the distant horizon; swiftly it comes up, gathers into its vast ocean the small streams and driblets of flight; it approaches the mighty host and is the mightier—devours, absorbs it—and, sailing grandly on, the vast accumulated multitude seems now to make the very air, and be, itself, the sky."[9]

One day Selous watched a group of starlings fuss and fight with woodpeckers, each peeved at the other in competition for tree holes. He became lost in the joy of watching the disputes and petty thievery, remarking that the disagreements found resolution in the way that only nature can—and that perhaps the rest of us might take heed. "What a bore all morality seems, as one watches them. How tiresome it is to be human and high in the scale! Those who would shake off the cobwebs—who are tired of teachings and preachings and heavy-high novelings, who would see things anew, and not mattering, rubbing their eyes and forgetting their dignities, missions, destinies, virtues, and the rest of it—let them come and watch a colony of starlings at work in a gravel-pit."[10]

He took particular delight in the starlings' flocks and their nightly gatherings. He watched them closely and often, sometimes trying to pinpoint the precise moment when they would decide, collectively, to move from daytime feeding in the orchards to their nighttime tree roosts, typically as the afternoon turned into evening. At first a few began the ritual of gathering in the trees, while others watched and waited. Some wheeled around in the air, deciding whether the time was right, Selous noticed, and then a silence was broken by vague babbling, "a felt atmosphere of song."

One by one, or sometimes in a group, they would head toward the forest, some flying, some walking through the fields, and then there would be an eruption: "All at once, with a whirring hurricane of wings, they rise like brown spray of the earth, and, mounting above one of the highest elms, come sweeping suddenly down upon it, in the most violent and erratic manner, whizzing and zigzagging about from side to side, as they descend, and making a loud rushing sound with the wings." Once they settled, he said, the tree was alive with song, "as though from some great natural musical box, a mighty volume of sound that is like the splash of waters mingled with a sharper, steelier note—the dropping of innumerable needles on a marble floor."

More birds arrived from the surrounding countryside, joining the clamor and then rising again into the air, stunning local villagers. "From tree to field, from earth to sky, again, is flung and whirled about the brown, throbbing mantle of life and joy; nature grows glad with sound and commotion, children shout and clap their hands, old village women run to the doors of cottages to gaze and wonder—the starlings make them young. Blessed, harmless community! The men are out, no guns are there, it is like the golden age."[11]

Yet others took to the sky:

> And now, more faster than the eye can take it in, band grows upon band, the air is heavy with the ceaseless sweep of pinions, till, glinting and gleaming, their weary wayfaring turned to swiftest arrows of triumphant flight—toil become ecstasy, prose an epic song—with rush and roar of wings, with a mighty commotion, all sweep together, into one enormous cloud. And still they circle; now dense like a polished roof, now disseminated like the meshes of some vast all-heaven-sweeping net, now darkening, now flashing out a million rays of light, wheeling, rending, tearing, darting, crossing, and piercing one another—a madness in the sky.[12]

I pictured Selous, seated on a rock in the field, notebook in hand but pencil idle, agog at what was unfolding. "All is the starlings' now; they are no more birds, but part of elemental nature, a thing affecting and controlling other things."

It seemed he had never-ending appetite for watching the starlings near his village, especially as great flocks separated into distinct smaller groups and merged again in a shape-shifting mass, "becoming with the quickness of light, a balloon, an oil-flask, a long, narrow, myriad-winged serpent, rapidly thridding the air, a comet with tail streaked suddenly out, or a huge scarf flung above the sky in folds and shimmers."[13]

He marveled at how the birds knew when to zig and zag, operating as, it seemed, a single body. "Seeing it produces a strange sense of contrast, which has a strange effect upon one. It is order in disorder, the utmost perfection of the one in the very height of the other—a governed chaos. Every element of confusion is there, but there is no confusion."[14]

Future scientists would have to decipher how they did it, he said, "but to me it appears to offer as good evidence for some sort of thought-transference."[15] His theories about a psychic connection between these birds were ignored, only to be revisited decades later as scientists probed the mystery of murmurations through math, data analysis, and complex computer modeling.

During his time in Icklingham, Selous was well aware of the bitter feelings that more and more farmers harbored for starlings in Great Britain. He noted "the usual shoutings of the Philistines—their cries for blood and fierce instigations to slaughter."

Selous was sickened and enraged by those who wanted to kill starlings in the name of saving some of the profits at their orchards. When the *Daily Telegraph* ran a story with a call to wipe out local starlings, Selous submitted an indignant response, railing against the extermination of the birds, partly because so many of their mysteries had yet to be solved. "These gatherings present us with a problem of deep interest. Who can explain those varied, ordered movements, those marvelous aerial maneuverings . . . vast crowded masses of birds, flying thick as flakes in a snow-storm? Is there nothing to observe here, nothing to study? Are we only to disturb and destroy?"[16]

His plea felt as if it were meant to echo across the Atlantic and find purchase in the debate beginning to rage in North America, not just about Schieffelin's starlings and their future on the continent but indeed about which wild animals deserved to exist at all.

9

The Skies Transformed

WHEN EUROPEAN STARLINGS FIRST ARRIVED ON THE NORTH American continent, it wasn't the same place it had been even a century before. Where there once had been millions of hulking, free-roaming bison, just a few hundred remnants huddled in a scattering of herds. Bears, wolves, and jaguars had been decimated, as had sea otters and fur seals on the West Coast, white-tailed deer, beavers, and countless other wild creatures. Within a geological blink of an eye—over the course of 150 to 200 years—profiteering, settlement, greed, fear, guns, and lawlessness had obliterated so much of America's teeming suite of animals that it was barely recognizable as the place it once was. Some of the species survived this initial slaughter, but others disappeared forever into the abyss, including Merriam's elk in the Southwest, monk seals in the Caribbean, and the Rocky Mountain locust in the West.

It wasn't just land that was transformed by extinction. The skies changed too.

The most startling loss overhead was the disappearance of the passenger pigeon, a bird once so stupendously abundant that its giant flying flocks sometimes took days to pass by a single spot. It's estimated that there were three billion passenger pigeons in North America during the nineteenth century, making it the most plentiful wild bird on the planet. They were found primarily east of the Rocky Mountains. When they arrived in vast undulating flocks, the skies darkened, a thunderous noise erupted overhead, and some people dropped to their knees in prayer, sure that the end times had finally arrived.

But any reverence for these birds proved short-lived. They were slaughtered on a colossal scale with whatever tools were at hand. John James Audubon reported the arrival and hunt of the pigeons like this:

> The noise which they made, though yet distant, reminded me of a hard gale at sea. . . . As the birds arrived and passed over me, I felt a current of air that surprised me. Thousands were soon knocked down by the pole-men. The birds continued to pour in. The fires were lighted, and a magnificent, as well as wonderful and almost terrifying, sight, presented itself. The Pigeons, arriving by thousands, alighted everywhere, one above another, until solid masses . . . were formed on the branches all around. Here and there the perches gave way under the weight with a crash, and, falling to the ground, destroyed hundreds of the birds beneath. . . . It was a scene of uproar and confusion. I found it quite useless to speak, or even to shout to those persons who were nearest to me. Even the reports of the guns were seldom heard, and I was made aware of the firing only by seeing the shooters reloading.[1]

In a letter from the 1870s, a Massachusetts man described a killing field of passenger pigeons: "They used to bait the ground and then in some way throw a net over the birds, and then they killed them by crushing in the heads with the ball of the thumb. And George Day told me that when the thumb got too lame, Hosmer would crush them with his eye teeth."[2]

Passenger pigeons became a staple in everyday diets, including in pigeon pies, often with the bird's feet laid atop the crust. New York City's famed Delmonico's served passenger pigeons alongside ham and truffles to its upscale customers. They were lucrative too. Bird feathers on women's hats became a widespread phenomenon for several decades, driving up prices and fueling insatiable demand. Sport shooters also paid handsomely for live pigeons, and the net-bearing pigeon chasers were only too happy to oblige by capturing them and shipping them by rail to shooting contests.

The wave of killing, alongside lost habitat as fields and forests were converted to farms and cities, squeezed hard. Within decades, the number of passenger pigeons plummeted from several billion to just a handful. The last passenger pigeon, a female named Martha, died in 1914 at the Cincinnati Zoo.

The passenger pigeons' demise wasn't the only stunning extinction among America's birds. Heath hens, portly grouse that lived in coastal areas from Maine to the Carolinas, survived for thousands of years until the 1800s, when hunting and fire suppression altered their habitat and diminished their numbers. By the end of the century there were none left on the mainland, with the remainder surviving for a time on the island of Martha's Vineyard. Despite rescue attempts, the last one died in 1932.

North America's only native member of the parrot family, the colorful Carolina parakeet, once flew freely in the forests of the eastern United States, a remarkable sight for anyone lucky enough to spot one. The last one seen in the wild was early in the twentieth century. The final captive Carolina parakeet died in 1918. Their demise, as with the others, was the result of multiple factors, including habitat conversion for human use, feather collection for women's hats, and extermination by those who deemed them a pest.

Great auks, large flightless birds that bred on remote rocky islands along the North Atlantic coast, were considered extinct by the mid-1800s. They had been chased into oblivion by those who disregarded their awkward beauty in pursuit of something more savage and short-sighted, slaughtering them for food and bait.

Although there were pockets of people who tried to stave off the extinction of these native birds, they were vastly outnumbered by profiteers, indiscriminate killers, and those ignorant of or indifferent to the quiet plight of wildlife disappearing around them. Compounding the problem were those keen on importing birds for their own amusement and willing to refill the skies with birds of their choosing.

IT'S NO STRETCH TO SAY THAT DURING THE SECOND HALF OF the nineteenth century and the first quarter of the twentieth, more nonnative birds were brought to America and set free than at any time before or since. Although house sparrows and starlings made the most headlines, there were scores of others.

In a 1928 report for the U.S. government, the ornithologist John C. Phillips attempted to record all the bird introductions in the United States but eventually threw up his hands in frustration. "The early history of the introduction of foreign birds into this country is mostly clothed in darkness," he concluded. "The records of many attempts, if such there were, have long been buried in the back numbers of local newspapers, and if any experiment was unsuccessful, it was soon forgotten. Hence, one trying to get an accurate idea of what has happened soon realizes that he is following a hopeless quest."[3]

Phillips, a Harvard-trained doctor, spent two years in the medical corps during World War I before returning to some of his first loves: birds, science, and the outdoors. He's credited with naming at least a dozen bird species or subspecies and published more than two hundred reports, ranging from simple observations to historical research, genetics, and conservation. When it came time to try to understand how many birds had been brought to the United States and released, he wanted to be thorough. The task was far from simple because he was trying to understand a great, scattered, and decentralized effort across time and space. The human motivations, he soon realized, were just as varied as the bird species involved: hunting clubs looking for more and better birds to shoot, farmers in search of birds to gobble insects plaguing their crops, newly arrived immigrants sentimental for the birds of home, and even those looking to fill the air with the melodies of more bird songs. Sometimes it was clubs—sportsmen's clubs, songbird clubs, or acclimatization societies—that paid for birds to be brought across the ocean and released. Other times individuals ponied up to set birds free. Whatever the case, little thought was given to the long-term repercussions of these introductions, and few bothered to consult with scientists who might have encouraged them to consider more than their own desires. "A great number of real biological experiments," perhaps as many as 90 percent, were "unrecorded and almost unknown to the ornithologist," Phillips said.[4]

Most of the early introductions involved birds that could be hunted. Colonists imported pheasants and partridges for sport shooting. George Washington noted in his journals that his friend the Marquis de Lafayette gifted several types of pheasants and some French partridges to Mount Vernon in 1786. Game birds were introduced in 1790 to the New Jersey estate of Richard Bache, a son-in-law of Ben Franklin. Those kinds of quiet, small-scale releases continued for decades and accelerated over time, including quail, pheasants, grouse, and other birds preferred by hunters, and these species are still around today.

Nearly from the start, there were two competing schools of thought when it came to introducing foreign birds to American soil. The first, promoted by conservationists like Joseph Grinnell and others, held that it's more important to protect native birds and not tempt fate by introducing nonnative birds that might harbor diseases or unleash some other unforeseen consequence. "The other school," Phillips recorded with some disdain, "would bring in anything from a button quail to an ostrich without any regard to the general suitability of the species."[5]

In that second category was Thomas Woodcock, head of the Natural History Society of Brooklyn, who received several crates full of birds from England—European goldfinches, bullfinches, linnets, and others—and released them in the city in 1846. Skylarks, a particular favorite among bird introducers, were released in Wilmington, Delaware; Portland, Oregon; British Columbia; and on the outskirts of Cincinnati, Brooklyn, and Long Island. Some 200 were liberated in New Jersey in 1880; another 150 near San Jose, California, in 1896; and 200 more in Santa Cruz County, California, a couple of decades later. By one estimate, as many as 7,000 skylarks were set free in New York between 1900 and 1915.

Most efforts ended in disappointment. If the birds didn't die in transit—sometimes half the flock was wiped out during the cross-ocean journey—they were often condemned to struggle after they were uncaged in a foreign land. In the wild, many were simply never seen again. Others were observed for a few months and then died. Still more seemed to gain an initial foothold, even living and breeding for twenty or so years in their new environment before petering out. European goldfinches, for example, proved to be more adaptable than some of their fellow foreign songbirds, while others, like Eurasian skylarks, had a tougher time. Ultimately, in this flurry of introductions, just a small percentage of bird species actually found their footing in North America, most notably pheasants and other game birds, along with house sparrows and starlings.

BIRDS WEREN'T THE ONLY NEWCOMERS. NEARLY 12 MILLION human immigrants arrived in the United States between 1870 and 1900, most of them from Germany, England, Ireland, and China. Another 14.5 million immigrants came between 1900 and 1920, often from southern and eastern Europe. Predictably, nativist movements flared up among those who claimed they were here first, ignoring Indigenous populations dating back thousands of years. Debate roiled over how to define this emerging American identity. Who belonged and who didn't? Who would find favor with the dominant established culture, and who would be marginalized? Who deserved protection and accommodation, and who would be despised and persecuted?

Some of the sentiments against human immigrants may also have been reflected in attitudes toward the newly arrived birds. House sparrows were among those where the comparisons found expression first, especially because of their reputation as an ugly and messy urban bird in the late 1800s. Many old-guard Americans turned up their noses as the nation's cities swelled with immigrants—and some offered similar disdain for the birds who came from afar. One overt example is Frank Bolles, a writer and secretary at Harvard University who split his time between Cambridge, Massachusetts, and rural New Hampshire. His father worked in President Grant's administration, and his family was well connected and deeply rooted in New England. He was part of an ornithological club and advocated for the protection of regional forests. And yet in an 1892 article for the *New England Magazine* that mostly lavished praise on warblers, owls, woodpeckers, and other native bird species, he couldn't resist closing his piece by inveighing against sparrows, city dwellers, and immigrants in the same breath.

"City-bred man without knowledge of lake and forest, mountain and ocean, is an inferior product of the race; but disagreeable as he is, the city-bred bird is worse," Bolles wrote. "The English sparrow stands to me as the feathered embodiment of those instincts and passions which belong to the lowest class of foreign immigrants. The Chicago anarchist, the New York rough, the Boston pugilist can all be identified in his turbulent and dirty society. He is a bird of the city, rich in city vices, expedients, and miseries."[6]

Starlings got much the same treatment as they became more established. The rhetoric around them, although sometimes more veiled, carried the same kind of invective hurled toward immigrants coming to America in later generations: accusations of dark and mysterious motives and unclean lifestyles, questions of legitimacy, worries over disease and the displacement of those already here. When, exactly, does a foreigner become a native? How many generations does it take? Or does it depend on whether they are viewed as useful and unobtrusive?

For decades, dichotomies abounded when it came to America's birds. Native species vanished, while foreign ones, a few at least, flourished. Sometimes the drama played out in nearly real time. "Affluent New Yorkers might have had pigeon pie on their plates in the 1870s, but they had live sparrows on their window ledges, under their eaves, in their backyards, in their streets, and in their parks—joined increasingly, after 1890, by starlings," the environmental historian Peter Coates wrote more than a century later.[7]

Soon enough, starlings were more than merely birds. They became objects of derision, flying things committing the unpardonable sin of eating crops meant for human appetites. They left messes in our streets and drove prized native birds out of their nests. Indeed, they'd become a problem to be solved, a war to be fought.

10

Appetites

EDWIN KALMBACH GREW UP IN A TIME BEFORE STARLINGS really took hold in America. Born in 1884, he spent his childhood in Grand Rapids and harbored a love for the outdoors—and birds. Not long after high school, he was hired as a taxidermist at a local science museum, and he soon moved up to become curator and assistant director. In the summer of 1907 he organized a two-month canoe expedition on Michigan's Grand River, where he and a high school student collected bird specimens, studied bird life, and took photos of the river valley. It was apparently the first-ever study of birds in that area and even yielded documentation of a few new species there. By 1910 he'd been hired by the U.S. Department of Agriculture's Bureau of Biological Survey in Washington DC (the predecessor to the U.S. Fish and Wildlife Service), launching a restless, groundbreaking scientific career that lasted until his 1954 retirement and beyond.[1]

Although much of his most noted work was around waterfowl (especially helping identify why hundreds of thousands of ducks were dying in the West), he never seemed to stray very far from starlings. His name shows up all over the federal government's work on *Sturnus vulgaris* for much of the twentieth century, from understanding their feeding habits and capturing them in church steeples to founding a brand-new research wing and providing his best advice for the public on how to live with these new neighbors. Kalmbach was mostly a self-taught biologist, and much of his research relied on time spent outdoors doing old-school, hands-on field work, the latest in a succession of ornithologists trying to unlock the secrets of the creatures overhead.

The early study of birds on the continent typically revolved around a simple question: Who lives here? Some of the first publications were surveys of particular areas, such as Mark Catesby's two-volume *The Natural History of Carolina, Florida and the Bahama Islands*, published in 1731–43, which included descriptions of 113 birds. William Bartram's work in the Southeast, *Travels*, added more details in 1791, and so did *Barton's Fragments of the Natural History of Pennsylvania*, by Benjamin Smith Barton, in 1799. *Birds of America*, a major study by the name most associated with birds in America, John James Audubon, was published between 1827 and 1838. It provided an expansive, colorful look at hundreds of bird species, along with information about their habitat, plumage, and habits.[2]

It wasn't until decades later, as more and more people attempted to domesticate the wild landscapes of America on a massive scale, that scientists turned their attention to the *economics* of birds. In particular, which birds were advantageous for farmers, and which were pests? More pointedly, which were worth leaving alone, and which should be shot?

Not long after he was hired at the Bureau of Biological Survey, Kalmbach spent a summer in Utah trying to figure out whether house sparrows were heroes or villains (per the many complaints) when it came to some of the local crops, especially alfalfa. He collected a number of stomachs of sparrows and other birds in the area and brought them back to DC, finding that despite their reputation as a nuisance, the sparrows were mostly feeding on weevils and other insects, which was actually doing the farmers a favor. He sent his findings to the head of the Utah game and fish department, who urged the public to rethink its stance on sparrows. "Birds should be protected and even the despised English sparrow should be thought kindly of while he is doing so much good," the game warden said. "Granted that he is a pest himself, and a robber and a thief, we can afford to feed him and to keep him around while he is working for us."[3]

By the second decade of the 1900s public perceptions had mostly soured on Schieffelin's starlings. People in the cities shook their heads at the nightly chattering and the messes left behind on streets and sidewalks. Rural dwellers harped on the birds' appetite for cherries and other agricultural items. More careful observers, though, noticed that starlings also ate a lot of irksome insects, including nonnative ones, that were pests to both people and crops. In some cases, the starlings seemed to do a better job of eating those bugs than the native birds. So, on balance, were we better off with starlings? That's what Kalmbach and fellow scientist Ira Noel Gabrielson aimed to find out.

Gabrielson, a couple of years younger than Kalmbach, was an ambitious scientist too, with an intense interest in both birds and insects. He eventually went on to become chief of the Bureau of Biological Survey and later the U.S. Fish and Wildlife Service. But before climbing the ranks, he and Kalmbach spent countless hours in the field and the lab trying to get to know the starlings in their midst. They divvied up the six states where starlings were common in 1916: Kalmbach got Pennsylvania, New Jersey, and most of New York, while Gabrielson took Massachusetts, Rhode Island, Connecticut, and Long Island, New York. They conducted continuous observations of starlings from April through October, the period when most of the public complaints came in. When they weren't in the field, they combed through the stomach contents of well over two thousand starlings, at that point likely the largest number used to investigate the eating habits of a single bird species.

The dead starlings came from all over the six states, some collected by Kalmbach and Gabrielson, others provided by students, government agents, farmers, and everyday citizens. Each stomach was sliced open so the researchers could meticulously pick through the bird's last meals. Bits of insects abounded: heads, wings, thoraxes, limbs, and more. Every morsel—be it insect or plant—was counted and classified, then laid out on a sheet of paper so the entire collection could be examined at once. Their painstaking research found that on average, about 41 percent of an adult starling's yearly diet consisted of insects. The proportion of insects peaked in the late summer and early fall to nearly 57 percent, and it fell in the winter, when there were fewer insects, to around 23 percent. The variety was also staggering. It seemed that starlings had been munching on a who's who of despised insects (at least, in some circles), including locusts, cutworms, grasshoppers, crickets, weevils, and beetles, some of which had been introduced into North America and were getting out of hand. There was also a scattering of bees, wasps, millipedes, spiders, and ants. The starling, they found, had few equals in the Northeast when it came to the sheer number and variety of insects consumed.

But what about crops? By then stories were told and retold about how starlings would ravage a fruit orchard in no time flat and leave a trail of destruction in their wake. One farmer in Connecticut said a flock of about three hundred starlings stripped an entire year's crop from a single cherry tree in less than fifteen minutes. Anecdotes were one thing, but data was another. Kalmbach and Gabrielson found that just 169 of the starling stomachs contained cherry bits and estimated that the fruit was less than 2 percent of the diet for all the starlings they examined. They drew a comparison between robins and starlings.

Robins were apt to feed in loose flocks, rarely congregating in great numbers but feasting on cherries constantly and producing a slow, steady drain of crops across a wide region. "On the other hand, starlings, the young of which are the chief offenders, frequently gather in large flocks, and, swooping down on a single tree, completely strip it of fruit while other trees in the neighborhood may remain untouched," Kalmbach and Gabrielson wrote in their 1921 report. "As a result, while practically every cherry grower complains of the robin, those who suffer from the more spectacular rise of the starling are much more bitter in their complaints."[4] So while not every farmer was hit by starlings, those who were felt especially stung.

The investigation went on to largely exonerate starlings from accusations of eating deep into crops like sweet corn. Hours of observation showed that the corn was often being eaten by the company that starlings kept, especially red-winged blackbirds and grackles. From the more than two thousand stomachs examined, Kalmbach and Gabrielson estimated that corn made up less than

1 percent of their annual diet. It was much the same for farm-raised apples, pears, peaches, and grapes. The starlings were more likely to be dining on wild fruits, like mulberries, blackberries, chokecherries, and elderberries—and mostly when insects were scarce. The research project, the most intensive of its kind to that point, found that starlings mostly ate bugs and wild fruit, while just 6 percent of their annual diet was cultivated crops. Perhaps the dark reputation that had dogged starlings was off the mark.

Even before the findings were officially made public, Kalmbach was making his case to the press that locals in DC ought to soften their view. "The starling is a desirable bird. It will be of especial value to the farmer as an insect eater," Kalmbach told the *Washington Times* in 1920, adding that it was more beneficial than robins and could be a boon for gardeners irritated by worms and grasshoppers.[5]

Kalmbach and Gabrielson's report, titled *Economic Value of the Starling in the United States*, was published in 1921 as the U.S. Department of Agriculture's Bulletin No. 868. It was an examination of nearly all that was known about starlings at the time, including what they ate and how they behaved. Curious readers could get a copy of the sixty-seven-page bulletin for 25 cents. The researchers noted that while most states with starlings had laws that lifted protections that were afforded other birds, perhaps it was worth rethinking that approach. They suggested a more nuanced management that would allow starlings to be killed if they were intensely ravaging crops but otherwise would leave them alone, save the occasional nonlethal effort to shoo them off a farm with loud noises. That way farmers could enjoy the benefits of these voracious insect eaters. "With its ready ability to adapt itself to new environments," they said, "the starling possesses almost unlimited capacity for good."[6]

Their nearly heroic assessment reminds me of something Pliny the Elder, the Roman naturalist and philosopher who died in the year 79, said about another species of starlings and insects centuries ago. When crop-hungry locusts approached, locals near Mount Cadmus (in modern-day Turkey) prayed to Jupiter, and soon rose-colored starlings arrived from over the horizon and feasted on the crunchy insects. "It is not known where they come from nor where they go when they depart, and they are never seen again except when their protection is needed," Pliny observed.[7]

The message from Kalmbach and Gabrielson, though, mostly fell on deaf ears. Any hopes they may have had that their study would definitively settle the starling question were dashed as *Sturnus vulgaris* continued its westward campaign.

THE ARRIVAL OF ANY SUDDEN EPIDEMIC OR OUTBREAK—BE it a disease or an avian influx—is bound to send certain people toward mathematical calculations. The British-Canadian naturalist and ornithologist Henry Mousley did some math of his own and figured that Eugene Schieffelin's starlings from Central Park had become millions, invading lands more than a hundred miles north of New York City, plus some that had already made it into Quebec. The birds were on their way, he noted in 1924, to becoming "more or less of a plague."[8]

Despite the research by Kalmbach and Gabrielson, the government sounded its own alarm. The U.S. Department of Agriculture issued a press release in the spring of 1925 announcing that starlings—sometimes just a few intrepid loners—had found their way to places like Athens, Georgia; Detroit; Milwaukee; and the banks of the Mississippi River near Baton Rouge, Louisiana. They'd reached Ohio, Alabama, and Ottawa, Canada, too. They were abundant across southern New England, southern New York, northern New Jersey, and eastern Pennsylvania. In some places, the starlings now outnumbered the house sparrows.[9]

In Orange, New Jersey, thousands of starlings and grackles often amassed in the early evenings at a single spot. "All night there is occasional noise and at dawn a grand murmuration as the birds depart, after disturbing many," one report said.[10]

The starlings' expansion was often haphazard and spotty. Congregations arrived suddenly in marshes or open land, where they stayed for a few days, weeks, or even months. Often one farmer's land was picked over, while a neighbor's land was spared. The noise could stretch on endlessly, only to abruptly drop away as the flock, responding to a signal only its members understood, moved on. Cities weren't immune. Thousands upon thousands of starlings showed up in Baltimore, where residents lost sleep amid the endless cacophony of those perched in shade trees and on building cornices. The police set traps for the birds and lit explosives, while the fire department trained water hoses on the roosts in vain. Irritation morphed into light panic over birds. Maryland's game warden called on citizens across the state to kill starlings, crows, and others "in any manner possible and at any time." There was even talk of enlisting the Boy Scouts to help with the slaughter.[11]

In the fall of 1929 the Associated Press ran a national story saying that starlings had arrived at the eastern edge of America's corn belt, including parts of Wisconsin. The development "adds a strident note to the birds' menacing cry of victory in an alien land. It removes all doubt that the starling is now a thoroughly entrenched member of the North American avifauna and however unwelcome its presence may be to many, it is here to stay."[12]

Around that same time, ornithologist May Thacher Cooke at the Bureau of Biological Survey estimated that the starlings' range had nearly doubled between 1924 and 1928. She, like most everyone else, was trying to come to grips with this new reality. Although the starling was a "natural vagrant" and a source of anxiety for many, Cooke remained hopeful that its effect on crops wouldn't be a bad as some predicted. And just maybe, she said, the wide expanses of the Great Plains and the towering heights of the Rocky Mountains would prove an impenetrable barrier to starlings' ability to reach the Pacific Ocean.[13]

BACK IN NEW YORK CITY, STARLINGS TOOK OVER ENTIRE blocks and buildings. By January 1933 one hundred thousand were roosting each night in the limestone facade of the Metropolitan Museum of Art, not far from where Eugene Schieffelin had opened his cages a few decades earlier.

The problem wasn't just the noise or the mess. The spectacle on Fifth Avenue now drew crowds of onlookers curious about this epic gathering of birds in one of the world's most urban areas. They watched at dusk as the flock ferried in and took over every windowsill, ledge, and horizontal space on the outside of the building, which stretched nearly four blocks. The starlings perched on the facade's sculptures and roof edge like tiny snipers. Soon the air crackled with calls. The *New York Times*, in a story headlined "Starling Invasion a Police Problem," reported, "The massive museum seems to sing with this great feathered chorus enveloping it, and at a distance the multitudinous chirping resembles the sweep of a thousand strings in a huge symphony orchestra."[14]

The starlings' music lasted through the night, keeping up the light sleepers in nearby apartments. Police complaints went nowhere—what could they do?—and the museum had no inclination to try to repel the birds from the building. The only relief came at dawn, when the birds peeled away toward the farm fields of Long Island. There they fed during the day, only to return to the museum at sundown in multitudinous black clouds of wings and feathers. So it went every winter for years. "Naturalists declare this vast conclave of birds to be one of the most curious phenomena ever witnessed in New York," said the *Times*.

It was perhaps no accident that the starling assemblies at the art museum were referred to as a symphony. Certainly, not everyone celebrated the din of the birds, but over the centuries there have been those who have turned their ears toward them and found the noise fascinating, even beautiful.

11

Sing a Song of Starlings

IN THE LAST DAYS OF AUTUMN 1920 ABOUT TWENTY STARlings spent much of their time on a quiet residential street in Brooklyn's Bay Ridge neighborhood, lounging on the interior branches of the linden trees, perching along telephone wires, and drinking melted snow in the streets. They were also singing. And while they did, someone went to a piano in one of the nearby houses and began plunking in search of the matching notes. One call had a distinct lilt, three G notes in a row followed by an F-sharp. The pitch was nearly perfect, and as the piano player strained to hear the formation of the starlings' songs, the bird notes seemed to lock into the organized structures of human musical forms. Sometimes they'd sing triads—chords of three notes sung simultaneously—that rang out in lovely harmony. At least ten different calls were ultimately identified and notated, and these included arpeggios, tetrachords, major and minor keys, even a diminished seventh chord.

"A remarkable thing in a bird!" wrote Marcia Brownell Bready, who analyzed the bird music transposed to piano in her 1929 book *The European Starling on His Westward Way*. "For days afterward, the recollection brought a smile of pleasure."[1]

Her book focused heavily on the starlings' habits and history before plunging headlong into their musical output. Starling songs weren't as brilliant or as loud as thrush songs, nor as variable as those of the mockingbird or thrasher, but they were strong when it came to clarity, pitch accuracy, and grasp of the diatonic scale.

Bready took pains to compare the musical notes of different birds' songs, pointing out the semitone intervals of the mourning dove, the progressive descent of the screech owl, and the ascending thirds of the red-winged blackbird. The starling was capable of hitting an impressive range of notes—about fifteen when taken to the piano, from the G below middle C to an octave above the F just above middle C. Bready found that each song had its own form, some with a strict rhythm and others with pauses and even key changes. Occasionally there were shades of an English ballad or something reminiscent of Johannes Brahms. Different songs were sung at different times of day, including busy calls in the morning and easy, languorous notes that fell with satisfaction in the afternoon. "The song . . . is of gentle singing quality, the chord form deliberate, the scale-form gliding," Bready said of the late-day tune, adding that it was a "soft but fascinating song, which seems to be simply the expression of an overflow of vigor or good spirits."[2]

I couldn't help but contrast Bready's delight and fascination with the starlings' songbook with the less-flattering reactions from others at the time. "Squeaks and gurgles, interspersed with pleasant musical notes," Alice E. Ball, a poet and teacher from Ohio, offered in her 1923 book, *Bird Biographies*. "A flock of starlings makes a great deal of noise."[3]

Bready wouldn't have it. She clearly relished identifying complicated musical phrases—tetrachords in the diatonic scale, transposition of notes to mark a key change, even implied harmony—spilling from the neighborhood starlings. "It is not too much to say that musically he is man's equal," Bready cooed, "in kind, if not in degree."[4]

Around the same time that the songs were being notated in Bay Ridge, a doctor and avid bird lover named Charles W. Townsend was watching—and listening to—the starlings in the coastal town of Ipswich, Massachusetts. This time, though, it wasn't original songs he was hearing but imitations of other birds. On a single spring day along the marshy shores, Townsend witnessed starlings perfectly mimicking robins, herring gulls, cowbirds, meadowlarks, wood-pewees, "bell-notes and scream" of blue jays, and finally, greater yellowlegs, a lanky shorebird. "When this last distinctive and familiar call came rolling forth, I instinctively turned and swept the marsh with my eyes, expecting to see the flashing white rump of this bird, but on turning back to the Starling, I saw his raised head and moving bill which revealed the mimic," Townsend told the attendees at the American Ornithologists' Union meeting in October 1923.[5]

A little more than a year later another doctor and ornithologist, Winsor M. Tyler, reported a similar experience in the same part of the state. He noted a group of starlings simulating other birds in rapid succession. Over the course of ten minutes, he said, he heard thirteen notes from ten different bird species: "the song of the phoebe, the whistle and the scatter call of the bobwhite, the *wee-chew, wee-chew* of the flicker, the song, nearly perfect, of the meadowlark, the sharp call-note and some vireo-like phrases of the purple finch, the wood pewee's peaceful whistle, the rolling *too-wheedle* of the blue jay, the two-note whistle of the chickadee, a note unmistakably that of the goldfinch, and the red-wing blackbird's cluck and the gurgling part of its song."[6] The scene was all the more remarkable because many of these imitated birds had been gone for weeks or even months, so the starlings had remembered their sounds and somehow called them all up from memory during that frenzied song session.

EUROPEAN STARLINGS COME FROM A LONG LINE OF MIMICS. The Sturnidae family includes more than a hundred types of starlings and mynas, many of which have a knack for imitating sounds. (Parrots, cockatoos, macaws, and similar mimicking birds are part of a different family, the Psittacidae.) Among the starling family, the European starling is one of the best at reproducing sounds. "It mimics so perfectly as often to deceive the most experienced ear," wrote Henry L. Saxby in *The Birds of Shetland*, his 1874 treatise on avian life in the far northern reaches of Scotland.[7]

Starlings were common in the Shetlands, with breeding sites including sea cliffs, turf and stone walls, stacks of peat, and even rabbit burrows, he said. They gathered in great groups, whether in shrubs or atop the islands' rocky stubble, sometimes moving into farmers' fields and even frequenting the carcasses of cows, sheep, and ponies. Saxby couldn't recommend them as a source of food for people, though, saying, "It would be difficult to find a more disgusting feeder." All the while, the starlings erupted in noise, often replicating the sounds of herring gulls, redshanks, plovers, whimbrels, curlews, and oystercatchers. Sometimes the starlings even resorted to their own songs. "Its own spring note is well known and appreciated, but it occasionally degenerates into a perfect medley of sounds, including the notes of other species which are decidedly unmusical," Saxby said.

In England, starlings had an expanded repertoire, with the calls of sparrows, yellow buntings, chaffinch, pheasants, woodpeckers, and blackbirds. "Besides these," the British-Canadian naturalist Henry Mousley wrote in a letter to Massachusetts's Charles Townsend, "he has been known to imitate perfectly a dog whistle, and the tinkle of a particular cycle bell, the latter so perfectly as to delude its hearers!"[8]

But its typical call tended not to inspire much romance in those trying to describe it. One British guide to birds from 1943 described the starling's song as "a lively rambling melody of throaty warbling, chirruping, clicking and gurgling notes interspersed with musical whistles and pervaded by a peculiar creaking quality."[9] In Australia, one starling observer described several starlings vocalizing at once: "One bird called plaintively as if in trouble. The call of another suggested the not-unpleasant sound of a good violinist tuning his instrument. In a third call we heard the slight crack of a child's whip, and in still another the harsher sound of a boy's rattle. Then we were startled to hear a spirited imitation of the blue wren's notes. But the leading note most characteristic of the starling, was one which suggested the breath drawn in with a musical sound."[10] Such vivid descriptions left curious researchers wanting to learn more.

In the springs of 1982 and 1983 Andrew M. Hindmarsh of the Edward Grey Institute of Field Ornithology took a boat to Fair Isle, one of the Shetland Islands, to collect audio recordings of hundreds of migrating starlings. The windswept island, about eighty miles off the northern edge of Scotland, was home to a few dozen people and thousands of birds, some living there most of the year and others just passing through. Hindmarsh not only collected starling calls in person but also was given twelve hours of starling recordings from the isle to analyze. The following year he published an exacting study of *Sturnus vulgaris* called "Vocal Mimicry in Starlings," an attempt to better understand the species' habits of melody and imitation. Each "song bout" usually lasted twenty to forty seconds, he said, and "begins with several calls and ends with a series of high-pitched screams. In the middle is the lively melody, frequently characterized by a rattling sound produced by the bill.... The imitations are generally very accurate, witnessed by the fact that most ornithologists can relate an anecdote about how they have been fooled by a starling."[11]

Hindmarsh found that the starlings on the isle made about 30 different mimicked calls with 239 variants, most of them imitating other wild birds on the isle, including curlews, lapwings, plovers, thrushes, and oystercatchers. On occasion the starlings would mimic the sound of a lamb or barking dog. Although the starlings' catalog wasn't as prolific as those of some other birds—marsh warblers had been known to imitate as many as seventy-six species, Hindmarsh recalled—they had a broad repertoire that relied on the natural sounds around them.

He offered a few theories about the starlings' behavior, including that imitating the calls of a competing species with aggressive or territorial calls might keep them away. They could also do the same in copying the sounds of a predatory bird, like a hawk. He also couldn't discount that the variety of songs could be intended to attract a mate or perhaps that the birds were making mistakes as they tried to expand the songbook, perhaps picking up tunes from other starlings that had been learned elsewhere. The simplest explanation, he ventured, was that the tunes of other birds were easy to learn, especially calls they heard every day from their neighbors like the curlews or oystercatchers. But it wasn't just birds that starlings loved to imitate.

IN VIENNA ON APRIL 12, 1784, WOLFGANG AMADEUS MOZART added his 453rd completed composition to his logbook, the lovely and lively Piano Concerto No. 17 in G. Satisfied, he was excited for its public debut in mid-June by one of his young students. The composition was something of a secret until then, and undoubtedly it ran through his head like a loop that was difficult to turn off. On May 27 Mozart went to a pet shop in town and, as the story goes, was whistling a phrase from his newest work. Imagine his surprise to hear the tune sung back to him nearly perfectly. The singer? Amazingly, a starling in the store had mimicked four bars of the song, hitting every note except at the end, when it changed two Gs to G-sharps, giving the end of the phrase a warbling, off-key effect. "Das war schön!" (That was beautiful!), Mozart said.[12]

Delighted, Mozart paid 34 kreutzer (around $10 in today's money) for the starling, noting the price and notation of the bird's song in his expense book, and went home with a new pet. Later accounts said the bird was named Star, but there's little concrete evidence of that. No matter, the starling became part of the daily life at the apartment that twenty-nine-year-old Mozart shared with his wife, Constanze, and their dog, Gauckerl. The starling doesn't appear much in the written documents left behind by Mozart, but he clearly developed a relationship with the bird as he composed his next works in Vienna. It's not hard to imagine that some of the starling's colorful songs may have woven themselves into his melodies in some way. Even if not, clearly a friendship had blossomed.

When the bird died three years later, the composer held an elaborate service for his little starling friend and wrote a poem for the occasion, which reads in part:

A little fool lies here
Whom I held dear—
A starling in the prime
Of his brief time,
Whose doom it was to drain
Death's bitter pain.
Thinking of this, my heart
Is riven apart.
Oh reader! Shed a tear,
You also, here.
He was not naughty, quite,
But gay and bright,
And under all his brag
A foolish wag.[13]

Centuries later, in 1990, scholars examining captive starlings' songs and voices—and their profound ability to mimic sounds—speculated that Mozart's first piece completed after the death of the starling, *A Musical Joke*, included hints of the bird in the notations, something they called "a vocal autograph" of the starling. In particular, the researchers cited the presence of "drawn-out, wandering phrases of uncertain structure" that were akin to starling vocalizations. What's more, they said, "the abrupt end, as if the instruments had simply ceased to work, has the signature of starlings written all over it."[14]

The study, which ran in *American Scientist*, was conducted by Meredith J. West, then a psychology professor at Indiana University, and Andrew King, research associate professor at Duke University. Not everyone has agreed with their conclusion, and West and King didn't pretend that Mozart's starling influenced the entirety of *A Musical Joke*. Some thought the musical piece, also referred to as K. 522, was a satirical take on other contemporary composers, perhaps a jab at what he viewed as subpar works. Although Mozart's true intentions aren't known, West and King said that "it is hard to avoid the conclusion that some of the fragments of K. 522 originated in Mozart's interactions with the starling during its three-year tenure. The completion of the work eight days after the bird's death might then have been motivated by Mozart's desire to fashion an appropriate musical farewell, a requiem of sorts for his avian friend."

THE GREEKS AND ROMANS KNEW THE EUROPEAN STARLING and its abilities well. Starlings were kept as caged birds in Greek houses as far back as the fifth century, and Aristotle sometimes used starlings as a way to compare sizes of other species. Romans kept European starlings in cages and taught them to speak. Pliny the Elder, writing in first-century Rome, observed starlings in captivity: "The young princess had a starling and also nightingales that were actually trained to talk Greek and Latin, and moreover practiced diligently and spoke new phrases every day, in still longer sentences. Birds are taught to talk in private where no other utterance can interrupt, with the trainer sitting by them to keep on repeating the words he wants retained, and coaxing them with morsels of food."[15]

In *A History of the Birds of Europe*, Henry Dresser noted a story about a starling in 1582 that sang and spoke in German and Polish and had even given itself a Polish name. Unsurprisingly, those skills have intrigued scientists for years.[16]

One of my favorite studies of starling sounds was outlined in the same 1990 paper where Meredith J. West and Andrew King examined Mozart's music and the influence of his pet starling. In it, they also recounted the decade-long linguistic exploits of fourteen starlings, most of which were hand-reared and then kept in people's homes. Some cohabitated with other birds; others were the only birds in the house. All were exposed to people and then, to varying degrees, they mimicked what they heard. Sometimes the starlings' words were immediately recognizable, and sometimes they were unintelligible. They also whistled tunes they heard and occasionally made up their own.

The whole thing seemed utterly charming and strange. The most often mimicked words were the birds' names, along with "Hi" and "Goodbye," "Give me a kiss," or "I think you're right." "One bird . . . frequently repeated, 'We'll see you later' and 'I'll see you soon.' The phrase was often shortened to 'We'll see,' sounding more like a parental ploy than an abbreviated farewell," the authors wrote. "Another bird often mimicked the phrase 'basic research' but mixed it with other phrases, as in 'Basic research, it's true, I guess that's right.'"[17]

The starlings' words and phrases were carefully noted and recorded again and again, utterances often incongruous with the situation. One of the "basic research" birds shrieked "Basic research!" when he had his head accidentally caught in a string. "Another screeched 'I have a question!' as she squirmed while being held to have her feet treated for an infection," West and King wrote. One picked up a phrase from a TV commercial: "Does Hammacher Schlemmer have a toll-free number?"

Much of the starlings' time was devoted to singing "rambling tunes" that were usually a mix of something they'd heard and something whose origin was a mystery. They liked to sing off-key and fracture the phrasing of the music in unexpected ways—at least, to our human ears. "One bird whistled the notes associated with the words 'Way down upon the Swa-,' never adding '-nee

River,' even after thousands of promptings," the authors said. "The phrase was often followed by a whistle of his own creation, then a fragment of 'The Star-Spangled Banner.'" Another bird took to whistling the opening melody of "I've Been Working on the Railroad" but with strong accents on the second line, like he was shouting, "All the livelong day!"[18]

The starlings also imitated intonation and even other human sounds, like lips smacking and throats clearing. One bird mimicked the sound of a person sniffing before saying "Hi," a development traced to its caregiver's ongoing allergies. Other starlings mimicked the sounds of keys rattling, dishes clinking together, and doors opening and closing. "Another chanted 'Defense!' when the television was on, a sound she apparently had acquired as she observed humans responding to basketball games," the researchers said.[19]

Today we know the range of starlings' imitations is wider than ever: cell phone rings, barking dogs, frogs, car alarms, baby squeals, dishes clinking, frog songs, and all manner of whistles, pops, and rattles. "How and why starlings incorporate such an elaborate repertoire of sounds into their vocal behavior has been the subject of a lot of research, with no single answer," said Rodney Sayler, a wildlife ecologist at Washington State University whose research includes avian behavior.[20]

It's possible the sounds are part of a complex interaction with other members of the flock, calling out their whereabouts, looking for love, or even trying to keep rivals away. Or it might simply be for fun. For now, they aren't saying.

12

Under Siege

AS THE WAVE OF STARLINGS ROLLED WEST ACROSS THE GREAT Plains in the 1920s and 1930s, newspapers and local bird journals lit up with reports. Some contained just the facts, while others vibrated with fear and anxiety.

Starlings' arrival in Nebraska proved a good example of how they were received. The *Nebraska Bird Review*, free to members of the Nebraska Ornithological Union and $1 per year for nonmembers, ran a litany of accounts as the birds arrived, first in a trickle and then a flood. The initial account came from a man named George Thiesen, who said he saw two pairs nesting in the cupola of his barn outside of Lincoln in 1930 and 1931. By the following summer "quite a flock" had returned to his farm. Thiesen tried but failed to capture a few of them alive. He finally shot one and brought it to the University of Nebraska Museum, where it was mounted. Farther south that same year, two starling pairs nested in the peak of a barn near Western, Nebraska, about thirty miles north of the Kansas border. A cat caught one of the young birds, but another was killed, dropped in alcohol, and also brought to the university.[1]

On the cusp of its arrival in droves, the starling often was assigned a mythical kind of status, like a mysterious and unseen enemy gathering at the far horizon. "A new and formidable pest for farmers of this section to combat is slowly working from east to west and has been reported in Iowa," the *Frontier* newspaper in O'Neill City, Nebraska, wrote in a front-page warning on February 22, 1934. "The starling is about the size of the blue bird, it is pugnacious, hardy, noisy and very prolific. It thrives everywhere, drives out other birds and will eat anything. Its song is entrancing."[2]

Reports soon poured in to the *Nebraska Bird Review* with a sense that the state was indeed under siege. One woman said her maid discovered a starling in the basement of their Lincoln home. They captured it and sent it to the zoology department at the University of Nebraska. "It is a mystery to me how the bird effected its entrance into the basement," Mrs. W. E. Barkeley wrote to the bird journal.[3]

Elsewhere, a trio of starlings grew to a flock of thirty to forty a few weeks later, one resident said, and another reported twenty-six starlings in a backyard, adding, "The bird seems to be increasing in this vicinity very rapidly." They came to downtown Omaha and homes on the outskirts, as well as to farms to the west, north, and south. Some starlings were met with shotgun blasts. Most simply hung around in the trees and shrubs and on the buildings, squawking and carrying on to the bewilderment of the locals. The first starling seen in the town of Wisner was apparently the one that traveled down the chimney of John Nuerenberger's home, through the furnace pipe, and into the soot chamber. It was caught, caged, and kept in the house as a pet.

Mrs. H. C. Johnson of Superior, Nebraska, said several hundred starlings were congregating in her barn "and proving to be considerable of a nuisance." Harry E. Weakly said that one evening in December 1937, he saw about twenty-five near North Platte. "The birds were very suspicious and hard to approach but I managed to get close enough to make certain of the identification," he said. He was able to shoot and kill one of the birds and sent it to the local high school's biology department. One man, a member of the Nebraska Game, Forestation and Parks Commission, said he counted twelve different starling flocks on a forty-mile drive one day in February 1938. "More than I have ever before seen in Nebraska," W. H. Lytle reported. "They seem to be coming into the state very rapidly now."[4]

Not every imagining of horror was off the mark. Soon after their arrival in the Great Plains came reports from Kansas, Nebraska, and Oklahoma that starlings were landing on the backs of cattle, apparently hunting for insects beneath the cows' skin. Ox warble flies laid their eggs on the legs of cows, and after hatching, the larvae would penetrate the skin and migrate around the body. Starlings hungry for those grubs would swarm cattle, perch on their backs, and probe beneath the skin with their beaks in search of food. The digging left behind open wounds, some said to be as large as a man's hat. The cows soon became fearful of the birds, unleashing surreal scenes in the pastures. "After having been once attacked, cattle will stampede upon the sudden appearance of starlings," Arthur L. Goodrich Jr. from Kansas State College wrote in a lengthy description in the April 1940 newsletter for the Kansas Entomological Society. "That the flying birds outstripped the running cattle and turned them by swooping in front of the racing cattle is surprising evidence of the resourcefulness of the starlings in their new undertaking."[5]

One stockman said a number of his cows had to be destroyed after they were injured when they slipped in icy pastures while trying to outrun the starlings. Cattlemen and veterinarians wrote to the scientists at Kansas State College after a bitterly cold winter when the birds' terrifying behavior left them confused and scared. "They pick for the 'worm or bug' which are in the hide of the critter, and don't stop at that but go on for 'blood and meat,'" one wrote. Another pleaded, "They are eating holes in the backs of our cattle. What is causing them to do this?"[6]

There were similar reports from Iowa, Texas, Oklahoma, and Nebraska. An entomologist in College Station, Texas, also passed along a strange anecdote from a conversation he'd had with a cattleman whose animals had been beset by hungry starlings pushing their bills into holes left by the grubs. Near the end of his story, the man added that "an animal that recently died from some cause had been entirely consumed by the starlings before he had an opportunity to dispose of it otherwise."[7] Each anecdote, specious or not, seemed to add to the legend of the fearsome starling.

AS THE 1930S DREW TO A CLOSE, AN APTLY NAMED ORNITHOLogist called Leonard Wing spent months compiling forty years of starling data from Christmas counts, the end-of-the-year bird survey carried out from coast to coast by volunteers and scientists alike. Wing, working from the State College of Washington in Pullman, wanted to get a snapshot of the status of starlings in America. Although they were spreading slightly more slowly than sparrows had decades earlier, Wing calculated that by 1940 starlings occupied 2.7 million miles in the United States, expanding from the East, through the plains, and shoving up against the Rocky Mountains. He braved a calculation, figuring there might be as many as seventy-eight million starlings in America, but realistically, the number probably didn't exceed fifty million. Wing guessed starlings made up about 1 percent of the total number of birds in the United States but believed the number was only going to increase in the years to come.[8]

A few other things were becoming clear about starlings, Wing said. They used advance teams to scout new territories, and once an area was deemed habitable, they arrived in bunches about five years later. They also migrated—generally moving from northeast to southwest—but the routes they took were not nearly as consistent as the migration pathways of native birds. Starlings were still feeling out their new territory and figuring out what worked best for them. Altitude did not seem to be a barrier to starling occupation, Wing said, noting they'd been seen in Jackson, Wyoming, at more than six thousand feet above sea level. Finally, it seemed obvious that starlings' expansion across the country tended to follow cities and valleys. They feasted in the fields and prairies by day and hunkered down in the safety of cities at night. People and their buildings attracted starlings, especially in the winter.

THAT FACT CAME AS NO SURPRISE TO ANYONE LIVING IN THE nation's capital. In late December 1927, as Washington DC was settling in for a quiet Christmas, four government men approached the massive U.S. Post Office headquarters on Pennsylvania Avenue. Nine stories tall and a block long, the Romanesque Revival building with its striking clock tower was one of the most attractive landmarks in the city. But lately, it had been beset with unwanted visitors. Each night about a thousand starlings crammed themselves into two ventilator units, happily perching where the building exhaled warm air.

Starlings first showed up in DC in the fall of 1914 as a few exploratory souls landed in the trees outside the U.S. Bureau of Fisheries office. By 1923 they were frequently seen on downtown window ledges and eaves, forming the nucleus of what would become a grand fall-to-spring infestation of tens of thousands that vexed the city for generations. Store owners complained about the noise and mess, and soon everyone in the nation's capital, including those in the highest reaches of government, seemed united in their irritation with the city's uninvited guests.

"Here their nocturnal squeals and chatterings reach the ears of the mighty and here also at times the voices of the mighty rise in protest," said a government report outlining the situation. "Here the shopper and the shop owner; the pedestrian and the autoist; the bird hater and even the bird lover periodically join the chorus of damnation. Even the staid ranks of profound ornithologists have echoed the song of lament."[9]

Local botanist and naturalist Donald C. Peattie noted how prevalent starlings were around Washington, especially in the sycamores lining Pennsylvania Avenue. The birds never failed to put on a show, whether it was displays of courage in stealing a morsel or the sheer outsized effect of their presence. "Chortling, wheezing, gabbling and whistling, they flock as dense as a black storm cloud, casting sarcastic comments on the passerby," he said. "Their song, if song it can be called, is laughably close to the chattering gossip of a hundred idle human tongues." Despite his interest, Peattie was clear that this new U.S. citizen was not particularly welcome. "Like the sparrow, he was brought as a friend and turns out to be a pest," he said.[10]

Experiments began in hopes of driving them away: explosions, traps, and even toxic gas directed at the roosts. Disappointed by the results, scientists at the U.S. Bureau of Biological Survey took up the case. Success in shooing the birds, they figured, depended on developing a better understanding of the starlings: their habits, their interests, and where they went when they left DC each spring.

That search for understanding is what had the four men from the Biological Survey climbing to the top of the post office building on December 21, 1927. Included among them was biologist Edwin Kalmbach, who had been involved in the study of more than two thousand starling stomachs over a decade earlier. The idea was to capture as many starlings as possible, affix numbered bands to their legs, release them, and then track their whereabouts over time. Starlings were known to hang out at the post office ventilators, so it seemed that it would be an easy enough task to snatch them there. The men, however, were quickly outsmarted. As they approached on the first night, they were backlit by the city lights, and most of the birds took off before the scientists could get close. They caught just a single starling that evening before calling it quits. Someone suggested they try the First Presbyterian Church on John Marshall Place, not far away, where thousands had made a home in the church tower.

The mess they found on the church landings was astonishing. In some places, the bird droppings were eight to ten inches deep. Being good and curious scientists, they examined these and found that a single quart contained more than 105 different foods, the majority of which were bugs, providing fresh insight into where the birds were traveling up and down the Potomac and in Virginia and Maryland.

The captures at the church began in early January 1928. When the crew arrived inside the tall church tower, they found starlings perched on ledges and cross braces. The birds had also stuffed themselves so tightly into a series of cavities in the tower that there scarcely seemed room for a single additional bird. And while the temperatures outside were frigid, the body heat of the birds kept the cavities toasty. "We ourselves were able to keep perfectly comfortable even though working bare-handed on cold nights, by frequently delving arms' length into the one of these cavities to drag forth a double handful of starlings," said Kalmbach, who led the capture project.[11]

The process was this: The men snatched fistfuls of starlings out of the cavities and stuffed them into gunnysacks in lots of forty or fifty birds per sack. They then took the sacks to a lower level, where each starling was inspected, a band was affixed, and then the bird was tossed from a window. Despite some rough treatment, the crew noted that just one bird died in the process. Most simply flew off into the winter's evening chill.

On the first night 317 were caught, banded, and released. A thousand got the same treatment ten days later, and another thousand in February. By the end of March more than forty-one hundred starlings had been banded. The following winter the men got nearly four hundred more starlings at another nearby church. The task then became waiting—waiting for those banded birds to fly to some distant place, die, and have their bodies found, and then for conscientious citizens to mail back the dead birds, or at least the bands and the locations where they were found. By the spring of 1931 about 120 bands had been returned, a disappointingly small data set but enough to draw some rough conclusions. When the starlings left DC, many of them went north, some as far as Vermont, upstate New York, and even Ontario. The majority, though, stayed within twenty miles of the city. Many of the starlings, the ornithologists concluded, could now be considered local birds.

For years after the capture project, starlings mostly avoided the church towers. Instead, they moved en masse to the old Land Office building on E Street, tucking themselves beneath the protected eaves and squatting on the upper-story window frames, sometimes with ten to fifteen birds crowded outside each window frame. Some forty-two hundred starlings, the bulk of the city's starling population, arrived at the Land Office building each night during the winter of 1929–30. The frustrating situation with the starlings at the church and the Post Office building had simply transferred to a new location.

At one point an experiment was tried at the Land Office building, where a wooden wedge cut at a forty-five-degree angle was placed on the outside window frames. The idea was that the starlings wouldn't be able to comfortably rest on the sloped wood and would find somewhere else to go. It worked, but within a couple of days of taking it down, the starlings were back, as frisky as ever. The next attempt was to fire an acetylene flash gun, essentially a device that creates an explosion about as loud as a shotgun accompanied by a great flash of light. It was set off over the course of several nights, targeting a portico clogged with starlings. The birds flushed at the commotion but slowly returned after the firing stopped a few nights later.

The situation took a turn for the absurd in February 1931. By then starlings were infesting every side of the Land Office building nightly and were spilling onto the U.S. Patent Office next door. The cacophony of more than six thousand starlings ramped up at twilight and seemed to drag on and on. A new, desperate plan was hatched. On February 9 four men went onto the roof of the Land Office, and another four posted atop the Patent Office building. Each had his own cat-o'-nine tails, a whip with wires attached to a short pole. Through the night, they lashed the ledges beneath the eaves like medieval combatants trying to slay a giant. On the ground, men hurled small rocks at the birds that couldn't be reached with the whips. It went on for two nights. On the third, the whips were exchanged for slingshots, and by the fourth night, somewhat miraculously, nearly all the starlings were gone. Some sort of victory could be claimed. "Now and then a small group will fly toward [the building] as if to alight on one of the ledges; they may even perch for a moment or two but it is not for long. As far as these two buildings are concerned the relief from the starlings has been complete," Kalmbach said in his report the following year.[12]

Of course, that didn't mean the problem was solved. About three thousand starlings relocated to a giant electrical sign advertising the popular Gayety burlesque theater on Ninth Street. The birds added a grim feel to the flashy sign that few seemed bothered by. Other starlings spread out in downtown Washington DC, and still more found their way to nearby Rock Creek Cemetery. "Not content to molest the living, Washington's pestiferous starlings now have launched a campaign to deface tombs of the dead," the city's *Evening Star* newspaper groused in a front-page story. The birds were not only despoiling gravestones but harassing cemetery visitors. Rifles and explosives were deployed "with indifferent success."[13]

Elsewhere in town, water was trained on the stubborn birds. "The fire department's hoses swept the birds into the air like so many bits of fluttering paper," the *New York Times* reported, "and the next day they came back for more."[14]

The starlings found their way over to the new Smithsonian Museum, where they were again targeted with an acetylene flash gun. Soon they were at the National Archives building, which was under construction. Building officials feared cleaning up the mess could cost as much as $20,000. This time crews lofted toy balloons to where the starlings were perching. It worked only until the balloons fell back to earth.[15]

As workers chased starlings around Washington DC, a debate

raged, and readers wrote to the *Evening Star* with their own ideas about what should be done. A sizable portion of the human populace sided with the starlings rather than their harassers. "No one seems to realize the starlings are valuable birds," one reader wrote in, with a reminder that the starlings were eating insects. "If these birds are observed in the daytime, instead of nighttime, their economic importance will be appreciated." Some said the starlings should simply be left alone. "In a certain sense they may be a nuisance, but what of it?" one woman asked. Far more annoying, she said, was the oil dripping onto the streets from cars, the pollution that clogged the air, and the burning of coal that smudged houses and apartment buildings. "These are surely a thousand times more of a nuisance than the roosting, for a few hours during the night, of the poor starlings who are not responsible for their existence." She added, "What if the streets do become dirty? They need cleaning anyway."[16]

Several suggested that the city build a series of boxes on pedestals high above Pennsylvania Avenue where the starlings could reliably hole up at night without making trouble for buildings or people. Use the boxes as advertising space, and heck, the bird boxes could even become an attraction for the out-of-towners. "Place them above the trees, with canvas boards beneath, flood them with light, and let the tourists come and look at them. That will be just one more sight to add to Washington."[17]

Another woman suggested that starling eggs be collected, boiled, and then replaced in the nests or exchanged for fake eggs. She also recommended building shelters in parks and then dividing starlings and pigeons by sex as a way to reduce breeding. "She also had some observations on birth control for humans and devoted 14 pages to her theories," the newspaper reported.[18]

One man offered the wild idea of a giant gadget like a windshield wiper that could sweep across the face of buildings and sweep any unwanted birds from their perches. "Such a device," he said, "would destroy the rest of the 'darling creatures' whenever they would seek to tuck their little feet under the warmth of their folded wings and would be worth the expense of wiring and the small amount of current necessary to operate it."[19]

Gigantic windshield wipers aside, Kalmbach and his team

remained on the hunt for viable options to deal with the DC starlings. After more than a decade of racing around town in search of the latest roost, they were ultimately left puzzled about why starlings could be permanently shooed from some spots but were only temporarily pushed out of others. The vagaries of the starlings' thoughts, strategies, and motivations remained highly elusive. At this point the starlings seemed to have the advantage in the contest against those thinking they might outsmart them. Could they ever be defeated? Should anyone even try?

Fig. 1. From its origins in parts of Europe, Asia, and North Africa, the common starling (*Sturnus vulgaris*) now occupies every continent except Antarctica. Scientists in 2021 estimated there were around 1.3 billion European starlings scattered across the planet. Only house sparrows are more numerous. Photo by Dan Vickers at Tallgrass Prairie National Preserve, courtesy of National Park Service.

Fig. 2. Eugene Schieffelin, bird lover and scion of a wealthy New York family, imported nearly two hundred starlings from Europe and released them in Central Park in 1889, 1890, and 1891. Although starlings were also released elsewhere in the country, Schieffelin's were the most important source of starlings in North America. They marched across the country and eventually numbered around two hundred million on the continent. Walter W. Spooner, ed., *Historic Families of America*, vol. 3 (New York: Historic Families Publishing Association, 1907).

Fig. 3. Even before starlings gained a foothold in North America, they elicited mixed feelings in Europe. Some early Romans revered them and tried to divine important information from their flight patterns. Elsewhere, farmers feared the arrival of tremendous, noisy flocks that would wreak havoc on orchards and crops. Lithograph by John Gould, Wikimedia Commons, https://commons.wikimedia.org/wiki/File:SturnusVulgarisGould.jpg.

Fig. 4. Starlings, seen here perched on the back of a bison, are one of nature's great survivors, capable of thriving in many types of environments. After they were introduced into North America, it didn't take long for them to move from the cities to more rural areas. Soon they arrived in the Great Plains and then reached the West Coast. Photo by Kim Acker, courtesy of National Park Service.

Figs. 5, 6. Few in the United States knew it at the time starlings were imported, but they are versatile songsters and incredible mimics. In the late autumn of 1920 the songs of a few starlings were notated in New York and later reprinted in Marcia Brownell Bready's 1929 book, *The European Starling on His Westward Way*. Photos by author.

Fig. 7. Starlings first appeared in Washington DC in the fall of 1914 and were soon roosting by the thousands in the city's churches, on building ledges, in trees, and anywhere else they could find. During the fall-to-spring occupations, shop owners complained about the mess, and politicians called for something to be done. Soon they showed up in cultural references, too, like this 1930 image by political cartoonist C. K. Berryman. National Archives.

Fig. 8. As starlings moved from east to west across the United States, their reputation soured among many in the public. Still, there were those who pushed back against the vilification of the newcomers, including biologist Rachel Carson. Her 1939 essay in *Nature* was titled "How About Citizen Papers for the Starling?" Smithsonian Institution.

STARLINGS — A NUISANCE EVERYWHERE

A PROVEN METHOD

Additional Satisfied Clients
I Ridded Of Starlings and Pigeons

AMERICAN FLETCHER NATIONAL BANK
Also 6 Other Buildings
MR. S. G. KASBERG, VICE-PRESIDENT
INDIANAPOLIS, INDIANA

VALLEY BANK BUILDING
GENERAL SERVICE ADMINISTRATION
MR. MARVIN NELSON
DES MOINES, IOWA

FEDERAL BUILDING
MR. T. B. HURST, BLDG. MANAGER — G.S.A.
LOUISVILLE, KENTUCKY

GRACE HOSPITAL
MR. MARVIN NICHOLS, ADMINISTRATOR
HUTCHINSON, KANSAS

PHOENIX TOWERS
MR. R. E. MALAN — MR. W. D. SMALL, MGRS.
PHOENIX, ARIZONA

MAHONING COUNTY COURT HOUSE
Also 7 Other Buildings
MR. JOHN PALERMO, COMMISSIONER
YOUNGSTOWN, OHIO

PAWNEE THEATRES
MR. AL McCLURE, MANAGER
WICHITA, KANSAS

BROOKOVER — 80 ACRE
CATTLE FEED YARDS
MR. STAN FANSHER, MANAGER
GARDEN CITY, KANSAS

STARLINGS AND PIGEONS SPREAD MANY
DISEASES, ALSO MESS UP YOUR PROPERTY

Write
OTTO D. STANDKE
"The Bird Man"
1305 BAKER AVENUE—GREAT BEND, KANSAS

OTTO D. STANDKE
IN HIS OFFICE AT

Fig. 9. The proliferation of starlings gave rise to a few Bird Men who made their living chasing starlings and other birds from buildings, neighborhoods, and city streets. Success was mixed and rarely permanent. Otto Standke, the Bird Man from Great Bend, Kansas, tried but failed to move a massive flock of starlings out of Mount Vernon, New York, in 1959. He went on to a well-publicized career as a starling chaser around the country. Photo courtesy of the *Great Bend Tribune*.

Fig. 10. One of the most popular ways to shoo starlings was blasting recordings of starlings in distress. Sometimes the noise was projected through speakers on mounted trucks, such as during the 1954 experiments to clear starlings from Philadelphia city hall. Other times, records of distress calls were given to people in the neighborhood, who would throw open their windows and turn up the volume in the early evenings as starlings arrived to roost in trees. Some of the records used during experiments in Denver in the early 1960s are in the archives at the U.S. Department of Agriculture's National Wildlife Research Center in Fort Collins, Colorado. USDA Public Archives. Photo by author.

Fig. 11. Eastern Airlines flight 375, a Lockheed Electra, ran into a flock of starlings just after takeoff on October 4, 1960, at Boston's Logan International Airport. Sixty-two people were killed, and it remains the deadliest collision between an airplane and birds in U.S. history. The crash prompted federal regulators and aviators to rethink the design of some airplane engines to better withstand bird strikes. Photo by the Civil Aeronautics Board.

Fig. 12. As starlings moved across the country, they often swarmed at large-scale cattle feeding operations, where they feasted on the cows' food. The issue was particularly pronounced in California and Texas. Poisoning and loud noises have been employed in hopes of keeping starlings away. Photo courtesy of USDA Wildlife Services.

Fig. 13. Part of the frustration over starlings' occupation of industrial agricultural operations is the mess they leave behind, like the droppings on this ranch trailer. Photo courtesy of USDA Wildlife Services.

Fig. 14. In response to public calls for controls on starling populations, particularly those roosting in trees around homes and military operations, government scientists began experimenting with new ways to kill the birds. One method was spraying a chemical on the birds that broke down the insulating oils on starlings' feathers and then hosing them with water, leaving them more vulnerable to the elements, especially cold temperatures. One agent, called PA-14, wiped out an estimated 11.4 million starlings at eighty different roosts during two decades of use. Photo courtesy of USDA Wildlife Services.

Fig. 15. Leg bands, like these used in Michigan, became a popular and widely used method to track the movements and migrations of birds. The technique was pioneered in 1899 by a Danish schoolteacher who tested it on a group of starlings in Europe, and soon it was a key tool for ornithologists in the United States. One starling banded in 1970 in Oregon traveled more than eight hundred miles. Photo by U.S. Fish and Wildlife Service.

Fig. 16. Skywatchers have long been fascinated by aerial murmurations of starlings, like this one photographed over a harvested cornfield near Starved Rock State Park in Illinois in the fall of 2007. The shifting, cloudlike formations of starlings seem to be equal parts math and poetry. For years, researchers have been trying to understand the purpose of the murmurations and how they might protect members of the flock from predators. Photo courtesy of Dan Dzurisin.

Fig. 17. Scientists have been studying how starlings were so successful in spreading across North America. Rapid genetic changes have helped them, as have the starling's larger brain size than other birds and an uncanny knack for survival in different types of environments. One study found that starling wings in the United States have become more rounded since their arrival, which may allow them to take off at steeper angles and gain a slight advantage in escaping predators. This photo is of a starling wing from Washington DC that's now part of the collection at the National Museum of Natural History. Smithsonian Institution.

Fig. 18. More than a century ago Josiah Wood Whymper created this wood engraving and watercolor of a starling in front of a church in England. Few birds elicit such a variety of responses from people as the reactions spurred by starlings: love, hate, frustration, bemusement, disgust, fascination. While starlings don't appear in the same numbers as they once did in the United States, they continue to be as divisive as ever, and it's likely they'll remain so. Courtesy of Wellcome Collection, London.

13

In Defense of Starlings

IN 1939 RACHEL CARSON'S CAREER AS A MARINE BIOLOGIST and writer was still in its early stages. For the previous few years, she'd been working for the U.S. Bureau of Fisheries, churning out pamphlets and stories meant to get the public excited about marine science. *Atlantic Monthly* had run an essay by her in the summer of 1937 that garnered her a book deal, which was published to acclaim in 1941 as *Under the Sea Wind*. She wrote several more books about the ocean before *Silent Spring*, her seminal 1962 work that exposed the dark sides of pesticides and helped launch the modern environmental movement. But for a time during the first half 1939, she took a break from writing about the sea to come to the defense of starlings.

By then the war on starlings was a common subject in newspapers, and Carson witnessed many of the machinations against the birds firsthand when she was in Washington DC for work. She'd seen their fall and winter congregations on buildings and had become a keen observer of their habits, noticing that they tended to arrive at their building roosts later on nice days in October than on days with a threat of inclement weather. And on clear mornings, they liked to leave their city roosts ten to fourteen minutes before sunrise—the early bird gets the beetle, as it were—while on cloudy days, they often stayed in their roosts just a bit longer, departing about five minutes before the sun came up. These observations weren't part of any official study of starlings; it was just that Carson's sharp eye and curious intellect seemed to never rest, and she came to appreciate the starlings in ways that ran counter to much of what official DC thought about them.

"One of the most fascinating sights I have ever witnessed is that of starlings in their aerial maneuverings before settling in for the night," Carson said in her essay "How about Citizenship Papers for the Starling?" published in *Nature Magazine* in the summer of 1939. "In mighty flocks, which grow moment by moment through the addition of new arrivals, they wheel and turn above the buildings, patterning the evening sky with intricate designs. Leaderless, apparently animated by the pure joy of flight, their performance is one of indescribable beauty."[1]

So while the wave of fury against starlings was rising in the eastern United States, Carson offered a poignant counterpoint. Rather than look at these birds with disgust, why not view them with wonder and empathy? Might we exercise a dash of tolerance for these cheerful, intelligent, and adaptable creatures?

In her essay, she noted how well the starling had done as a newcomer, adapting and then enduring decades of revulsion from the very species that had caged them up and brought them to America. "In spite of his remarkable success as a pioneer, the starling probably has fewer friends than almost any other creature with feathers," she wrote.[2]

But perhaps the unfavorable assessment had been hasty, especially in considering starlings' benefits to farmers. By one estimate, Carson noted, an individual starling might carry more than a hundred loads of "destructive insects" each day back to hungry offspring, while cramming its own stomach full of Japanese beetles, caterpillars, grasshoppers, weevils, and cutworms. Indeed, the starling had proved to be one of the best in combating pesky insects that terrorize crops and annoy people—better than sparrows, robins, red-winged blackbirds, grackles, and other birds, Carson said. On the other end of the starlings' appetites, they seemed to shoulder much blame when it came to devouring farmed fruits, but robin were also culprits in the orchard, and no one was chasing them around. "It has been charged that the starling harasses native birds and drives them out of their holes," she continued, "but it has recently been found that the gentle house wren is also a despoiler of nests, and who does not love the wren?"³

So much of the ire directed at starlings was over their tendency to congregate in cities like New York and DC during the winter. But, Carson argued, can you blame them? In their native areas of northern Europe, starlings fled during the colder months for southern Europe and the Mediterranean. On the foreign soil of America, with no information about where to migrate when the cold hit, starlings found that city buildings, with their sheltered porticos and enclosed church towers, provided warmth and cavities where they might cluster together and share body heat. These are survivors, Carson noted, doing what survivors do.

Carson knew starlings faced an uphill public relations battle. Scientists before her, like Edwin Kalmbach, had praised the starlings over the years in an attempt to sway those irritated by their flocks, droppings on coats and sidewalks, and feeding frenzies in which they stripped trees of fruit. It was a quieter, minority viewpoint that tended to occasionally find its footing, only to be overwhelmed again by public outrage in the aftermath of each new occupation. Even the U.S. Department of Agriculture tried to make a case for starlings, albeit a tepid one. Several years before Carson's plea, in the summer of 1931, the agency issued Farmers' Bulletin No. 1571, which reported Kalmbach's findings that most of the starlings' habits were either beneficial to people or at least economically neutral. "The time spent by starlings in destroying crops or in molesting other species of birds is extremely short compared with the endless hours they spend searching for insects or feeding on wild fruits," he wrote.⁴

It was a tough sell. Public attitudes tend to be shaped more by stories, anecdotes, lore, and legend than by scientific data. What would be more persuasive—a recounting of insects found in the stomachs of starlings or breathless accounts of cacophonous hordes that blackened the sky and seemed indifferent to every attempt to keep them away? It's always easier to assign sinister motives to those we don't understand, be they strangers in the human or the animal world. Still, when it comes to starlings, there have always been a few who have taken up the call to look past perceptions and assumptions.

Not long after Rachel Carson's piece, a first-person feature story in a prominent Washington DC magazine told the story of a family who adopted a starling that had shown up in their yard with a broken wing. Jesse James, as he was called, was doted on as a gregarious family pet while he convalesced. Neighbors stopped by to see him dancing on the curtain rod, imitating the dog and the family's hens, and sometimes calling out "Hello!" in a rusty squeak. He seemed determined to counteract some of the hostility his brethren had generated in the city over the previous decades. "The enemies of starlings claim that they are noisy, dirty and quarrelsome; they neglect to mention that where starlings congregate you'll find precious few tent caterpillars and Japanese beetles," the writer said. "I'd rather have starlings in the garden than bugs."[5]

Several years later, in 1947, the writer and naturalist Edwin Way Teale offered similar support for starlings. In the *Coronet*, a popular magazine in the same vein as *Reader's Digest*, Teale laid out his case in his essay "In Defense of the Pesky Starling," the same article where he offered a passing suggestion that Eugene Schieffelin had been inspired by William Shakespeare. Sure, the starlings were a mess and an annoyance—"the din is tremendous"—but one need only witness them gobbling lawn-destroying beetles and other bugs to see their value, he said. "In the course of time, nature's balance will be restored," Teale predicted. "The hosts of the starling will be reduced as the hosts of the English sparrow have been. And in the end, the starling—that noisy bird invader that has conquered a continent—may be regarded as an immigrant that has developed into a beneficial and valued citizen of the country."[6]

To the north, Canada's venerable newsmagazine *Maclean's* ran a long story in 1949 trying to discern the value of the country's most controversial bird, headlined "The Starling—Saint or Sinner?" This "angel in black" had recently become a nuisance in Canadian towns and cities, but on the whole, it was a helpful visitor for farmers seeking assistance with copious insects, the article said. One tobacco farmer in Ontario cringed at hearing gunners and policemen blazing away at roosting starlings in the city. "Every time one of those guns goes off," he said, "I can count on another hundred cutworms for next year. If those fellows could come out here during the day and see the flocks of starlings eating cutworms in my tobacco fields, they wouldn't be in such a rush to get out their shotguns."[7]

Many of the same methods used in the United States had been deployed north of the border to drive away the birds, including cannons, hoses, glue, molasses, colored balloons, and noisemakers. The result was much the same. *Maclean's* sided mostly with those who favored starlings, even as their neighbors seemed determined never to give up the fight. It mentioned an additional benefit: starlings provided at least a modicum of entertainment in their ability to mimic others. A nationwide Christmas radio broadcast had featured a captive starling from Ontario that squawked "I'm a naughty birdie," "You're crazy," and then finished the act by whistling several bars of "Home on the Range." "But saint or satan," the article said of the starling, "he has made more enemies than anything that bears feathers, including women's hats, and the campaign against him will go on as relentlessly—and futilely—as ever."[8]

Although the religious comparisons weren't exhausted, some writers eventually returned to the immigrant analogy, but this time infusing it with positivity, conferring on starlings some of the ideals associated with the most entrepreneurial aspects of America. After all, this was a place that supposedly celebrated the stories of immigrants arriving on its shores, working hard, and making something of themselves. A place that rewarded adaptability, tenacity, and chutzpah. Samuel Pickering Jr., a Connecticut birdwatcher and English teacher, had been watching starlings and other birds at his feeder and came to wonder, Who better exemplifies persistence and the pursuit of happiness?

"The starling is a true democrat," Pickering wrote in a 1981 essay titled "The Starling Is a Real American" in the *New York Times*. "Unlike the cardinal, which pecks and chooses and disdains suet, the starling eats everything and doesn't flutter on ceremony. To it, a bird is a bird. When the blue jay shows up loud and brassy, strutting around in its blue and white suit, the starling doesn't snub it like the precise goldfinch, which hurries away to more subdued companions. No, indeed, the starling just raises its head as if to say, 'How do you do?' and keeps eating. Unlike the red-winged blackbird with shiny epaulets or the evening grosbeak with his jeweled forehead, the starling doesn't dress up. Every day it wears ordinary work clothes and is proud to be a plain starling."

But success has a price, especially for those that aren't supposed to achieve it. "Like many other immigrants who left slower lives behind in the small villages of Europe and came to America to better their lots, the starling labored to achieve its present stature and in the process risked its health," Pickering wrote. "Success does not always make friends and the starling is not without enemies. It has alienated the Eastern bluebird aristocracy. For years Eastern bluebirds had the best nesting places in cavities in fence posts and old trees all to themselves. The starling's upward mobility upset the established pecking order and aroused animosities."

His essay made the case for fundamentally rethinking a bird that so many had come to despise and dismiss. Perhaps we should bestow on starlings some of the symbolism long foisted on bald eagles, he suggested. "The time has come to acknowledge that in the dawn's early light and the twilight's last gleaming one sees not the bald eagle, but the starling."[9]

SO WHAT OF THE STARLINGS' ACID REPUTATION AMONG SOME ornithologists and bird lovers, villainizing them as unrepentant homewreckers and nest stealers? Could it ever be rehabilitated?

All good science aims to challenge assumptions, especially those hardened into gospel. For starlings, one of the earliest and most oft-repeated criticisms was that they were driving native cavity-nesting birds out of their homes and having a detrimental effect on their populations. Ornithologists and bird lovers have fretted over the fate of bluebirds, flickers, purple martins, sapsuckers, woodpeckers, and swallows. Anecdotes piled up over the decades as aggressive starlings were observed usurping the homes of native birds, unceremoniously moving in and sending the original occupants off in search of new digs. Some bird lovers couldn't help but retaliate. During one seven-year period, more than four hundred starlings were caught and killed after invading a single nesting box set up for bluebirds.

More than a century after starlings' arrival, a University of California–Berkeley research zoologist went looking to see if all the worrying was really warranted. Walter D. Koenig examined nearly a century of data from Christmas bird counts, the Audubon Society's national bird survey conducted by volunteers every year, and thirty-one years of data from the North American Breeding Bird Survey, a large-scale survey by citizen scientists that provides population data and trends for more than five hundred species. In both of those data sets, Koenig examined population trends of twenty-seven native cavity-nesting bird species before and after the arrival of starlings.

If the stories were any indication, the data should have painted a worrisome picture for native birds. But Koenig, with his results published in 2003 in *Conservation Biology*, found something altogether different. Of the twenty-seven species he examined, seventeen of them (63 percent) showed no significant change in their population with the arrival and occupation of starlings. When it came to the remaining ten, only half showed declines that could potentially be attributed to starlings—but even those findings came with caveats. Those species were the American kestrel, yellow-bellied sapsucker, Nuttall's woodpecker, red-cockaded woodpecker, and eastern bluebird. Of those, the results were ambiguous for Nuttall's woodpecker and the kestrel: although one data set showed a decline, their numbers actually increased in the other. For the remaining three, each showed negative trends in the Christmas count data but "nonsignificant changes" in the other data. Among those species, only the sapsuckers seemed to face possible declines from starlings, but even that remained iffy.

There was more. Four species showed significant increases in populations after the arrival of starlings: purple martins, acorn woodpeckers, pileated woodpeckers, and red-bellied woodpeckers. Also, northern flickers significantly increased after starlings' arrival, but then their numbers dipped down closer to what they had been beforehand. Some of them may have been able to defend their nesting cavities when starlings came calling or may have nested successfully later in the season once the starlings had moved on.

"Taken as a whole," Koenig said, "these data fail to support the hypothesis that the North American starling invasion has severely affected native cavity-nesting birds."[10]

So what to make of this unexpected exoneration? Perhaps that while it's shocking to see starlings harassing these native birds and taking their homes, the damage is limited as far as populations are concerned. And maybe, Koenig said, there's now some coevolution going on. Woodpeckers, for example, might be forced to be more aggressive in defending their territory, and there could be a decline in the aggressiveness of starlings to claim those cavities. Ultimately, fewer nests might be usurped, he said, but that would be a difficult trend to detect.

Koenig knew the study wouldn't go over well among conservationists and others who viewed starlings as a scourge on native birds. Indeed, he'd experienced his own starling frustration firsthand in his earlier fieldwork on acorn woodpeckers when starlings took over the nests of his study site. But the overall data spoke for itself. "The evidence that this competition has led to significant population declines is pretty slim, at best," he said later. Still, the ornithologist couldn't bring himself to become a starling apologist: "I certainly can't say that's changed my attitude toward them."[11]

14

How to Kill a Starling

BY THE TIME EDWIN KALMBACH GOT AROUND TO WRITING his next major report, in 1940, on the best way to handle starlings, America had been doing its best for nearly a half century, with hardly anything to show for its efforts. Starlings' march across the continental United States had continued with only the slightest resistance. They were now breeding in North and South Dakota, Nebraska, and Kansas. During the nonbreeding season, they'd been spotted in Colorado, New Mexico, Montana, and Wyoming. They were destined to reach the West Coast in a few short years, completing their coast-to-coast domination of the continent faster than anyone could have ever imagined after their introduction in Central Park sixty years earlier.

Kalmbach, who had spearheaded the work on starlings in Washington DC and had been part of the team snatching and banding birds at the church tower, had spent thirty years as a biologist with the federal government. Much of his work was as a wildlife "food habits specialist," examining animal behavior and appetites, especially around farms, cities, and other human environments. His area of expertise had become pesky birds and providing specific ways to keep them away from crops and minimize the damage they caused to neighborhoods. Over the years, the government had published twenty-two of his bulletins addressing crows, magpies, and sparrows, as well as the spread of botulism through some species like ducks. Much of his work tried to determine the economic costs of troublesome birds, often by analyzing stomach contents, as he'd done with starlings years before. The next step was then to tailor treatments—either shooing birds away or killing them—at the places most vulnerable to large-scale bird occupations.

In the summer of 1940 Kalmbach was fifty-five and something of a wise elder among wildlife scientists, the kind of thorough investigator who had seen much and could be relied on to tackle the thorniest of problems. In August he was put in charge of a new wildlife research lab in Denver that combined his food habits work with that of other scientists looking for new ways to kill rodents, predators, and any other species that weren't deemed valuable. The government research went back to at least 1887, when Kalmbach's predecessors had devised ways to wipe out coyotes, jackrabbits, prairie dogs, bobcats, and other "nuisance" species. The methods embraced both the modern and the medieval: guns, traps, poisons, and even introduced viruses.

As the leader of the new lab and its twelve full-time scientists, Kalmbach faced the task of providing some fresh advice about dealing with devilish European starlings, both the great hordes that descended on crops and the stubborn urban visitors crowding onto building ledges, church steeples, and sidewalk trees. An artist and a poker player, Kalmbach was also something of a diplomat. In his resulting report, Wildlife Leaflet 172, which came out in December 1940, Kalmbach struck a note similar to his earlier take on *Sturnus vulgaris*. He acknowledged that while many hated the birds for their droppings, noise, and harms to some crops, starlings were an advantageous neighbor when it came to eating crop-destroying insects like grasshoppers, weevils, and certain beetles.

So rather than trying for the blanket destruction of starlings, Kalmbach offered a list of ways to cope with them, some more lethal than others. His ideas included draping nets over crops and church steeples, severely trimming trees where starlings roost, and creating loud noises to drive them from a particular spot. Those without access to those tools could try simply grabbing the base of a tree where starlings were roosting and giving it a good shake. It was a rudimentary, if highly localized, solution but likely cathartic for the most frustrated. People could ring bells, employ slingshots, use flashing lights, or shake pebbles in a can to scare off starlings, though those methods had met with only moderate success, he said. Where flocks were feeding on the ground, traps could be set—baited with stale bread, garbage, or old fruit—and starlings could be gathered up and dispatched. Kalmbach mentioned the cat-o'-nine tails deployed at the White House. And why not try air guns and fire hoses? "One often finds the birds outlasting the fire fighters, who may have more important calls for their services," Kalmbach noted.[1]

He also offered a more festive approach that had been used in Washington DC: hydrogen-filled balloons on strings that workers could raise and lower among the trees where starlings were roosting. Or the balloons could simply be attached to buildings, where they would sway menacingly in the breeze. "This inoffensive method has worked to advantage in the vicinity of hotels, where frightening by noisy measures would be objectionable," he said.[2]

For the most desperate, he offered more lethal suggestions but first cautioned soberly against wanton killing: "Wholesale destruction of the starling population as a means of reducing the nuisance of objectionable roosts has been suggested, but its practicability has yet to be demonstrated. In the present state of our knowledge and experience the benefits accruing from attempts at wholesale destruction appear to be restricted chiefly to the immediate vicinity of the roosts attacked."[3]

With that in mind, he said shooting was, of course, an option. The shotgun was a handy tool for frightening and killing starlings in rural areas. The prospect was trickier and more frowned upon in urban environments, he noted. That said, cities like Baltimore and Wilmington, Delaware, had become desperate enough to occasionally rope off city blocks so shooters could blast away at the birds. The dead starlings were given to the needy for food. Kalmbach recommended double-barreled, 12-gauge shotguns loaded with no. 7½ or 8 shot. The technique was most effective when multiple shooters worked in unison after dark when the starlings had settled into the trees for a night's rest. It would likely take several nights of shooting to achieve any real progress, Kalmbach said. "Unless large and dense concentrations can be attacked, the cost per bird killed, in labor and ammunition, mounts rapidly," he noted. "Under what appeared to be very favorable conditions, [ornithologist Lawrence] Hicks in Ohio was able to kill, in 14 well-spaced attempts, more than 4,000 birds at an ammunition cost of $1.34 a thousand."[4]

He mentioned offhandedly that one unintended consequence of shooting too many birds in rural areas was that those that escaped the bullets would simply move to the cities, where gunning was more difficult and the birds would potentially find safer havens. So, he seemed to suggest, *if you think you can overcome these rascals with guns, proceed with care.*

Extraordinarily, chemical weapons had also been investigated, Kalmbach said. In the 1920s researchers with the Biological Survey teamed up with the War Department's Chemical Warfare Service to conduct a series of experiments about whether gases deployed in the First World War could be effective in controlling starlings. They tried six different gases, with mixed results. The main problem was that when chemical was released in a high enough concentration to kill a large number of birds, it also became a health threat to people and livestock. Wind was also an issue, as was the actual delivery of the weapon. Sometimes the hissing sound of the poison being released was enough to drive birds away before the gas ever reached their lungs. The experiments continued, though. In the winter of 1935–36 crews sprayed hydrocyanic acid where starlings were roosting in tall building porticoes and outcroppings around Washington DC. Even in calm conditions, the gas clouds were quickly dispersed and diluted "to a point where asphyxiation of the birds on ledges 50 to 60 feet above the ground was irregular and uncertain." The result simply wasn't worth the cost and the labor, Kalmbach said. The crews tried spraying the gas in the trees that were packed with starlings, but the sound rousted the birds before the lethal effects could take hold. At other buildings, they tried sprinkling calcium-cyanide dust, which produces hydrocyanic acid gas when released in a humid atmosphere. This time they also brought an electric blower to penetrate hard-to-reach roosts. They had slightly more success there, but again, the cost and labor made this method "of doubtful utility."[5]

Where weapons failed, Kalmbach suggested that perhaps architecture could prevail. He said it was time to consider different designs for America's buildings, especially ones without the ledges, nooks, and slopes that were irresistible for roosting starlings. It was a small but remarkable suggestion: admitting defeat against starlings and saying people should now construct different kinds of buildings as a way to cohabitate with them. No more ledges, deep-set windows, or bold-relief ornamentation. "Probably no type of architecture lends itself to the needs of roosting birds better than the classic Grecian, with its deeply carved pediments, sheltered porticoes, and abundant columns, from the simple Doric to the highly ornate Corinthian," Kalmbach said. Clearly, he'd spent some time craning his neck to see where the starlings wedged themselves in urban environments. He went so far as to suggest the optimal angle for ledges to keep the starlings away: forty-five degrees and very smooth.[6]

So those were his recommendations. If anyone reading his report was looking for a surefire solution, they were left wanting. Sure, try nets and traps and noises and guns and poison gas, but don't expect to make much headway except for driving flocks from one spot to another, as they were sure to land somewhere else. The conundrum of the starlings, now nearly a half century into a forever war, endured.

WHILE GOVERNMENT SCIENTISTS LIKE EDWIN KALMBACH struggled to articulate a strategy for keeping starlings away, more people started trying their own methods, some clearly in the throes of desperation. In Yellow Springs, Ohio, for example, an ornithologist at Antioch College used a trained barn owl to kill twenty-four starlings that had taken to roosting in the towers of the college's administration building.[7]

Not far away, in Lima, Ohio, the local chamber of commerce hosted a "bird blitz," where the local Rifle and Revolver Club killed about three thousand starlings over three weeks. In one night alone, under police supervision, the group claimed it went through seven boxes of 12-gauge shotgun shells and collected 121 pounds of dead starlings.[8]

In Rome, New York, a man thought that fake owls might be a good deterrent for starlings—especially the individual birds that attempted to protect their families from this nighttime predator—so he rigged up a system to make them pay dearly. "I mounted a stuffed owl in a wire cage which was electrified. The first night that the cage was put at the top of a pine tree, four or five of the starlings were killed and from that time on the starling trouble disappeared in my yard," said the man, who wrote to the federal government about his experiment. "Before that I had had several thousand of them in my yard every night."[9]

A janitor in Milwaukee smeared grease all over the grillwork of the windows at a courthouse, "so alighting birds skidded, did flip-flops and generally lost interest in the building." In Rochester, New York, a radio truck equipped with an "experimental high frequency wave transmitter" was set up near a flock of starlings. "The new weapon sends out audio waves, which are above the human hearing range and so selective that only the starlings are affected by them," one report claimed, adding that it was successful in temporarily pushing the birds to another spot.[10]

Blunt force was the preferred approach for an Illinois farmer tired of having his corn crop decimated by starlings. He and others filled gallon pails with bait and a stick of dynamite and hung them in the trees where thousands of starlings perched each night. When the birds settled on the branches, a detonator set off the improvised explosives. "The ground was covered with dead starlings," a local report said.[11]

A chemist at Syracuse University named Benjamin Burtt built a set of large traps lined with chicken wire and put them on a few buildings downtown in the mid-1960s. Each was baited with grain and decoy birds. Starlings landed on the perches and slipped into the cages through tight, funnel-like openings. Once inside, they couldn't get out. In the winter of 1964–65 Burtt's traps captured 55,000 starlings, which were gassed and thrown away. In Waterbury, Connecticut, where about 250,000 starlings were gathering outside city hall, twelve police officers showed up at 3 a.m. on a December morning and fired no. 8 buckshot into the trees. About 2,000 birds fell dead, but the operation was halted after neighbors complained. Car windshields and sidewalks remained slick with smelly starling droppings. "You'd have to be crazy to love those birds," one city hall worker said. Talk turned to conducting another late-night raid, this time with a fire hose. Someone else suggested they just hang dirty diapers from the tree and otherwise leave the birds alone.[12]

A decades-long battle in downtown Muncie, Indiana—including shotguns, fireworks, and all the other usual weapons and accompanying carnage—yielded few results but a surplus of vexation. Local officials, fed up with the mess and smell at the county courthouse, where starlings had taken over, eventually tore the old building down, built a new one, and hoped for the best.[13]

Years later a Pennsylvania writer named Charles Fergus documented his neighborhood's occupation by starlings, which eventually evolved into a personal battle that unfolded in his own backyard. "The starling swept across America because it found an ecological niche unoccupied by a native bird—an incongruous, twentieth-century niche that included superhighways, bridge beams, building ledges, garbage dumps, grain fields, hog troughs, and the back yards of split-level homes," Fergus said in a collection of essays published as *The Wingless Crow*. "It is not a handsome bird. Its black feathers reflect blue and purple and green. Its long bill is shaped like a pair of needlenose pliers. Its tail is stubby, its wings triangular, its body chunky. In flight, it looks like a cigar with wings."[14]

He continued: "Not surprisingly, starlings are mistrustful of man. You cannot walk up to one and brain it with a rock, as you might a robin. To explore this wariness, I decided to conduct an unscientific experiment." One March morning a dozen or so starlings were perched on the bare branches of a walnut tree in his backyard. He went to his bathroom with his .22, opened the window, and took a shot. He killed one starling, but the rest took off, flew in a large circle, and landed again in the tree's branches. He shot another, and the rest flew away "for parts unknown." The starlings were back the next day, and again Fergus went to the bathroom window with his gun. As soon as he lifted the sash, the birds took off. They were in the tree later that day, and Fergus took a shot again, this time from a different part of the house. On his next attempt, the birds flew away at the first sound of him opening the door to the outside. "After that, the only way I could get a shot was to sneak out while a truck was rumbling past on the highway, tiptoe to the corner, kneel, and ease my rifle around," he wrote. "It didn't seem worth the trouble."[15]

MEANWHILE, IN WASHINGTON DC, DESPITE THE GOVERNment's new reports, the pressure mounted to do something, *anything*, about the hordes of starlings. The city, however, remained divided about whether the approach should be to kill the starlings or simply do what was needed to move them away from roosting in the worst possible spots.

Many ideas surfaced. In January 1948 William Xanten, the city's superintendent of sanitation, was so distressed that he bought twenty-five pounds of itching powder from a company in California and announced plans to release it on the starlings through the units used to spray insect-killing DDT. A local policeman said he was developing a mechanized red-tailed hawk that could be mounted near a roost. The fake bird could then flap its wings and wheel about like an oscillating fan, he explained, and maybe even make noise. The policeman envisioned mass-producing the hawks and taking care of the city's starling problem for good. Downtown, the owner of Hamilton National Bank covered all his building's ledges with sheet metal set at a forty-five-degree angle. Any starling attempting to land, he reported, skidded off into the street. Five other buildings were soon outfitted the same way. Someone else suggested putting out baited trays lined with oil so the birds would step in the goo and take it home with them, eventually coating their eggs and rendering them unable to hatch.[16]

Traps and shotguns were also discussed. Fake owls and more balloons and even rubber snakes were deployed. Live electrified wires were strung across the Corinthian columns at the Capitol and on other buildings. The starling siege continued for years, for decades. At the White House in the early 1960s, building crews tried some of the latest anti-starling technology: blasting them with recorded calls of injured starlings in hopes of keeping them off the lawn. The birds had taken up outside John F. Kennedy's bedroom, making such a racket that his staff called on the National Park Service for help with the recordings. After the maintenance crew played the calls for a couple of days from a tree outside the White House, Kennedy spokesman Pierre Salinger had a prolonged exchange with the press corps about it. At one point, Salinger had the recording played for the curious reporters. "Through the press office came the hideous 'aarwk, aaarwk, aarwk' sounds of a starling in real trouble," the *Boston Globe* reported in a front-page story on November 10, 1962.[17]

The reporters followed up with a series of questions, including where the recorded distress call had originated: "I suppose," Salinger shouted over the noise, "an ornithologist at some time made a recording. You don't have to physically distress a starling."[18] The peppering continued, according to a transcript in the *Globe* story.

Q. What is the starling in distress over?

Salinger: I don't know.

Q. Are similar records being played in other places around town?

Salinger: I don't know. I can only speak for the White House.

Q. Is that a single starling?

Salinger: A single starling.

Q. It sounds like a statement made in California.

Salinger: Let's turn off the starling, please. It is distressing me.

Distress or not, they might've finally been onto something.[19]

15

Blast 'Em with Starling Calls

FOR FIVE NIGHTS IN THE WINTER OF 1954, TWO TRUCKS SPENT hours slowly driving around the outside of Philadelphia city hall blasting the jarring screech of unhappy starlings from industrial-strength speakers mounted on the back. For the neighbors, it was an absurd sight born of an absurd situation. For those operating the trucks, it was the chance to test a hypothesis about how best to keep starlings at bay.

Starlings had swarmed the ornate building for fifteen years or so, typically showing up in early November, raising a ruckus throughout the winter, and leaving in early April. City hall, which covered an entire block and rose seven stories, was irresistible for starlings, with its towering cupola, ledges, hundreds of sculptures, and ornamental trim—all perfect nooks for resting and roosting. When the starlings were in town on winter nights, the majestic building in the middle of the city was overrun with the wings, feathers, and chatter of fifty thousand restive birds. "As darkness falls, they flock into the courtyard and onto the faces," one report said. "An observer on the roof in their midst has the impression of being in a black blizzard."[1]

City officials, frustrated with years of failure at trying to keep the birds off the building, agreed to be part of an experiment being run by a husband-and-wife team, Hubert and Mable Frings, along with Joseph Jumber of Pennsylvania State University. Hubert Frings and Jumber, who both worked in the zoology department at the university, had been looking for a solution to the starling problem that brought tens of thousands of the birds each winter to the university's hometown of State College, Pennsylvania.

One night they went to investigate starlings roosting at a barn outside of town. They caught a few and held them upside down by the legs. The birds let out a terrifying shriek, and the other starlings took off in a panic. Frings and Jumber were intrigued. Later they caught several more wild starlings and gave them the same treatment, holding them by their legs and shaking them roughly. This time, when the birds let out their piercing cries, a reel-to-reel recorder the team had set up caught the sounds on tape, usually about twenty seconds at a time. They collected enough to create an hour-long recording of nonstop starling distress calls.

In August 1953 the Frings and Jumber set out to test their theory that the recorded calls might serve as a repellent for starlings. The first stop was a quarter-mile stretch of a street in State College where some twenty thousand starlings perched each night in the Norway maples and elms. On the first try, they played the recording through loudspeakers beneath four trees. They were delighted to see the starlings cleared from the trees at the end of the second night. They cleared ten more trees during three more nights of blasting the starling cries—but noted that many of the birds simply relocated to nearby trees. The basic idea was there, though. Broadcasting the calls for about thirty minutes before sunset and another thirty minutes after sunset seemed to keep the starlings out of the targeted trees. If the starlings could be kept at bay as evening fell, the trees would likely remain unoccupied for the rest of the night, since starlings tended to avoid flying in the dark.

Satisfied, the Frings and Jumber scaled up. They piled the tape recorder, a thirty-watt amplifier, and some large speakers into the back of a truck and spent four nights driving around several blocks of town where starlings had been roosting. "We next tried to clear a whole town," with similar success, the pair said in a report that later appeared in the journal *Science*.[2]

Twenty miles to the east, in the small town of Millheim, Pennsylvania, starlings outnumbered people by about eight to one. At the height of the summer visits, there were about twelve thousand starlings divided between two massive roosts in town. For this job, the Frings and Jumber set up two trucks—each loudly playing the recordings—and stationed observers on top of buildings in the middle of town to track the starlings' movements. After three nights of chasing the birds around, fewer than one hundred were left in town. But again, many simply relocated, this time to a nearby woodlot. Still, the scientists declared success, noting that the birds didn't return to town for the rest of the year.

For the recordings to penetrate leafy branches and scare off the starlings, they had to be played at a harsh volume, around 120 decibels when measured from a meter away. That's about as loud as a chainsaw. "It is disturbing to man if heard near the horns but is not troublesome when directed upward through mobile speakers," the pair wrote. They also experimented with recordings of captive-raised starlings in distress, but these didn't have the same effect on roosting starlings. The distress calls of the wild starlings didn't do much to drive off other bird species that often roosted with starlings, including grackles and robins.[3]

Armed with fresh intelligence from their work in Millheim and State College, the Frings and Jumber traveled to Philadelphia, ready to attempt their grandest experiment yet. When it was built at the turn of the century, city hall was one of the largest and tallest free-standing masonry buildings, with a towering spire reaching more than five hundred feet and a vast interior courtyard. It was elegant, powerful, sprawling, and ideal for birds looking for a perch, especially in the recesses that shielded them from wintry winds. The Frings and Jumber spent a few days in Philadelphia just observing the building and the starlings, trying to understand the occupation and devise a plan. City hall "is made-to-order for starlings, and they show their appreciation by roosting on it in immense numbers," they observed.[4]

On the first night in late January 1954 they employed three sound trucks equipped with trumpet-style speakers and amplifiers. One truck was parked in the courtyard, and the other two slowly circled city hall on the streets outside. As night fell and the starlings began to settle on the building, the distress calls were played at maximum volume from all the trucks for about forty minutes. "Even to the human ear, the sound quality left much to be desired," the scientists said in their report. Inside the courtyard, the noise was horrendous as it bounced off the stone walls. The maximum volume distorted and muddied the recording to the point where it wasn't even recognizable as a bird call. Outside on the street, the sounds ricocheted and pierced the heart of the city, creating a dizzying din that was shrill and awful to listen to. Some of the birds at the upper reaches of city hall flew away, but as darkness set in, most of the starlings remained in their usual places. "It was clear that noise alone was ineffective; the sound had to be recognizable," the scientists said.[5]

They returned the next night with a new sound system, one that could play the recordings at top volume without sacrificing the quality. They blasted the call for more than an hour, and many of the birds flew wildly away and then mostly kept out of the courtyard perches. All told, Frings and Jumber figured there were about five thousand fewer starlings in the courtyard than the night before. Not bad. On the other hand, most of the birds driven from the courtyard were huddled in recesses near the large tower, tucked into nooks where the speakers couldn't be directed. The trucks outside the building didn't seem to budge many birds at all. The following night was much the same.

On the fourth night the crew mounted bullhorns on the roof in hopes of laying down a "wall of sound" to wash over the top of the building while the trucks did their best from below. "The flocks seemingly exploded as the sound was turned on them," they said. The starlings were held at bay until it was nearly dark. Some of the more cunning birds found their way to "sound shadows," crannies and ledges where the noise was blocked by the architecture of the building.[6]

After about an hour of the noise attack, the biggest flock was reduced to about 25 percent of its size, and nearly all the remaining birds were in places where the sound couldn't reach. The scientists, deciding that they'd need more speakers and other equipment to properly clear city hall of starlings, held off on the following night so they could observe whether starlings would return or if the trauma of the previous nights would be enough to keep them away. As darkness fell, the starlings returned to the building in copious numbers, as if they'd never been assaulted by the speakers the previous nights. Dejected, the Frings and Jumber packed up their things and left.

Their experiments continued elsewhere, including in Easton, Pennsylvania, and Rochester, New York. They weren't all a failure, by any means. In fact, at four of the seven locations during 1953 and 1954, nearly all the starlings were temporarily cleared from particular areas using the trucks and speakers. But it was neither simple nor foolproof—the adventure in Philadelphia had made that clear. Beating the starlings with noise required plenty of equipment, repeated treatments, the right recordings, and a working knowledge of the birds' behavior, especially the need to turn on the speakers before the starlings settled in for the night. Any success was typically local, a simple measure of relocating the birds—and the problem—somewhere else.

And there was one more factor to consider, Frings and Jumber said: the human one. Not surprisingly, each experiment created a fuss in the neighborhood where it took place. Publicly broadcasting the piercing cries of distressed animals at a blistering volume was sure to cause controversy and disruption. Although in the Rochester neighborhood, for instance, a majority of people wanted to be rid of the starlings there was "a vociferous minority" that wanted to see the treatment fail. "In part this was due to an understandable desire to see anyone fail in what they had tried too long without success to do," the scientists said.[7]

In Mount Vernon, New York, noise experiments created a circus-like atmosphere as the public and the media showed up in droves. Before work could get underway, impatient news photographers implored neighbors to bang on the trees with poles in hopes of flushing the birds so they could get a dramatic shot. Frings and Jumber tried to do their work more secretively the following nights, but that was difficult to do given the noisy ruckus they created. On the third night, the full-volume speakers drove a flock of starlings from one neighborhood, but when the birds landed in another, residents were waiting with guns and began shooting. It was a fiasco.

To boot, the starlings brought out cultural differences among people—some wanted the birds simply gone; others argued they had a right to exist where they were. Much of it depended on where people lived, how they related to wildlife, and how far they were willing to battle birds in pursuit of relief. Attitudes weren't uniform, so the public's reaction to the noise experiments were apt to be as complex as the treatment itself.

IN 1962 RESIDENTS OF A LEAFY NEIGHBORHOOD IN DENVER found themselves at the end of their rope. Thousands of starlings had set up in rows of Chinese and American elms, raining down feathers and droppings on sidewalks, cars, and patio furniture. Bodies of birds that had keeled over in the roost littered the ground. The noise, especially the incessant chattering starting at dawn, drove many residents to the brink. Some furiously trimmed branches to eliminate the roosts, and others sawed down entire trees, exchanging shade for the hope of peace and quiet. City crews came out with some of the usual tactics, like floating helium balloons above the birds and firing lit sparklers into the trees after dark. Some of the birds were momentarily irritated enough to move a few blocks away, but most ultimately stayed put.

By then officials at the federal wildlife research center in Colorado were well aware of the Frings' experiments in Pennsylvania and New York with starling distress calls. Perhaps it was time to give it a shot in Denver. Researchers tracked down a Denver radio DJ named Bill Pierson (who later became a well-known TV weatherman). Pierson helped them put bird distress calls on records, long players that spun at 33 RPM. Side 1 featured twelve stirring minutes of unbroken, panicked starling sounds that set one's teeth on edge. Side 2 was more of a greatest-hits collection: thirty-second clips, played continuously, of upset and frightened starlings, grackles, and robins.

In the fall of 1964 the records were passed out to thirty-one homes in the neighborhood, each bearing a white label with orange lettering and the slightly whimsical Signal Broadcast Productions, Inc. logo of Pierson's new bird sounds company. While some wanted to start playing the records right away, others took convincing to be part of such an usual community experiment. "Several residents were willing to play records only after they learned they would not be alone in the dispersal effort," researchers said.[8]

On the first appointed night, with more than four thousand

starlings and grackles perched high in the elms, eight neighbors pulled their record players and speakers outside to play the records they'd been given. Five other homes opened their windows and cranked up the calls on their home systems. The researchers, too, set up their own mobile record player at the edge of the roost. Again and again, the calls went out into the cool night.

The experiment went on for several evenings. By the fourth most of the starlings were gone, save for a few stragglers and some confused newcomers. A small group returned on the tenth night but were inspired to leave when four neighbors dropped the needle on Pierson's records and laid on the volume. The starlings stayed away for the rest of the fall, and neighbors were happy to finally get some decent sleep.

As ever with starlings, the story had no tidy ending. Neighbors about five blocks to the east soon gave the word that a thousand or more of the freshly displaced starlings and grackles had suddenly shown up.

The next year the starlings were back in full force in both neighborhoods, including one where more than ten thousand starlings and grackles started clustering in the summer. More records were given out, and more distress calls were blasted night after night. The noise served its purpose. Birds directly exposed moved along, and when some returned, often another brief discharge of the calls sent them on their way again. To the extent that the researchers could track where the starlings ultimately went, they found the birds in quieter rural areas outside the city, hunkered down and out of earshot of the disturbing calls they'd been subjected to.

The experiment provided enough success to seed a new business. As Pierson worked his way deeper into the TV and radio news business in Denver, he developed a bustling side gig selling his bird-call records, especially to those looking to get rid of starlings. Over the next fifteen or so years, his records were sent all over the country, including to a steel mill in West Virginia, pecan growers in Oklahoma, a pine grove in Maryland where bird droppings were several inches deep, and about a hundred police departments desperately flummoxed by bird roosts. Pierson's records were a pleasant surprise for a man named Don Seufert in Huntingburg, Indiana, as twenty thousand starlings and blackbirds harassed the furniture factory where he was head of security. "We tried everything, even shooting close to 2,000 birds a night for a week, but it did no good," he said in a 1977 news story. "There were more birds the next night." Seufert started spinning Pierson's records each evening, and the birds were gone in three weeks. He made it a habit to fire up the record every year at the onset of spring.[9]

16

Darkness in the Golden State

BY THE TIME ALFRED HITCHCOCK'S THRILLER *THE BIRDS* CAME out in the spring of 1963—a film built on an inexplicable but terrifying invasion of birds in the Northern California town of Bodega Bay—the Golden State was trying to unravel a real-life avian mystery of its own: what to do about the millions of dark, hungry, and enigmatic starlings that had crossed the Great Plains, swept over the Rocky Mountains, and were now occupying California from north to south.

It wasn't some theoretical exercise. While other places contended with noise and nuisance, in California the starlings had become so abundant that they were threatening to put a dent in the state's immensely lucrative agricultural empire, not to mention its burgeoning network of industrial-scale cattle feedlots. In the decades after the first few exploratory starlings showed up and subsequently ushered in vast clouds of their relatives, California became one of the nation's leading laboratories in the war on starlings. Who won? It depends on whom you ask.

The first official word of starlings in California came quietly, while most of the world was focused on the Second World War, in the form of a short paragraph in the *Condor*, a long-running ornithological journal initially established to cover bird issues in California and the West. In the March–April 1942 issue, naturalist Stanley Jewett, based in Portland, Oregon, said he'd gotten a report that about forty starlings had been spotted near the town of Tulelake, California, close to the state's northern border. Jewett had been sent a dead adult male starling that had been caught in the area on January 10 at 11:30 a.m. "So far as is known, this is the first record of the European Starling in the state of California," Jewett said.[1]

The first breeding pair of starlings in California was reported in 1949, and soon the floodgates opened. Within a decade, they were spotted from Imperial and San Diego Counties in the south to Modoc County in the north and points between. Hundreds had become thousands and eventually millions. As reported in the *Proceedings of the 2nd Vertebrate Pest Conference*, "Even though the starling may be unwanted in California, it is now here and there is little chance of extirpating it."[2]

In the summer, the resident population of starlings feasted on grapes, figs, olives, pears, and cherries. In the winter, migrating starlings arrived in great swarms at cattle feedlots, hog farms, and other places where foods such as grains were left in the open. Rather than roosting in cities like their eastern cousins, California's starlings tended to flock to rural spots, especially dense groves of trees and marshy areas within easy reach of agricultural operations. Vilification followed. "The starling, by all accounts, is the orneriest, cussedest and in every way the most disagreeable and destructive bird known to man," said a 1960 news story. Officials girded for a full-fledged occupation and a coming "war on starlings" in California. "We can expect the explosion here any time," said Conrad Schilling, a Fresno-based entomologist working for the state.[3]

Troubling reports came in. A single visit by a flock of starlings caused $10,000 in losses at a twenty-seven-acre vineyard. A stockman said starlings were ravaging the barley and corn used to fatten up his cows, leaving behind such a mess that they were costing him $1,000 a day. Another said he watched a huge group of starlings eat twelve tons of discarded french fries, sometimes used as food for cows, in a single day. By 1965 it was estimated there were about twenty-four million starlings in California, about an eighth of the total number in the United States and Canada. "They have only begun their buildup," Maurice Peterson, dean of agriculture at the University of California, warned darkly.[4]

Government agents knew of the failed efforts elsewhere in the country to get rid of starlings and that they couldn't afford to repeat mistakes or be complacent. "Every attack must be calculated on the latest information concerning the nesting and feeding habits of the starling," said Schilling, the Fresno entomologist, who helped poison nineteen thousand starlings in his county in just a few hours. "We can't go on what was successful last year, or even last month. It must be on the basis of what they're eating now and doing now."[5]

State and federal officials teamed up on an ambitious new project called the Starling Control Committee, pooling their money and resources for a series of years-long experiments aimed at limiting the damage of this new bird. California would become a giant laboratory to find the best ways to harass and kill starlings—something akin to the Manhattan Project but for nuisance birds. Noisemakers and flashing lights had their place, but this new enemy required more. Wide-scale poisoning wasn't off the table, nor were experiments with sleep deprivation, disruption of starling reproductive cycles, and even radiation. No state had ever done so much over such a long time to combat a single bird species.

The largest and deadliest operation was at a cattle feedlot operation in Solano County between San Francisco and Sacramento, not far from the mouth of the Sacramento River. During the winter of 1965–66, a series of wooden bait stations were set up and filled with food sprinkled with a powdery, yellow poison called DRC-1339, a slow-acting chemical that attacked the kidneys of birds, killing them in a few hours or days, depending on the dose. The technical name was derived from the federal Denver Research Center (DRC) and the fact that this was the 1,339th chemical tested there. The chemical, soon known as Starlicide, had been developed in a partnership between the federal government and Ralston-Purina, the company mostly known for pet foods. In lab tests, the poison kicked in within about four hours, making the starlings listless, groggy, and uninterested in food or water. It was unclear if they were in pain, scientists said, but the starlings didn't spasm or cry out or convulse. "The birds die a seemingly quiet death," federal reviewers later said.[6]

Starlicide had been employed before but never at the scale used that winter in Solano County, where poisoned bait was brought in by the truckload, including more than three tons of toxic barley, three tons of treated rolled milo, and more than two tons of poisoned raisins. Early in the experiment, the birds at the feedlot were mostly blackbirds and cowbirds. Within weeks of eating the poison, the population of 5,000 was reduced to about 250. The real test came a few weeks later, when millions of migrating starlings descended on the feedlot. They had no trouble finding the deadly bait, feasting madly for days at the troughs. At first several thousand dead starlings were found at the site, but as inspectors expanded their search area, the death count grew exponentially. Investigators estimated that at one roost in a swampy area west of the feedlot, more than two million starlings had died from the poison. The carnage was even more significant at nearby Sherman Island, where a population of five million starlings had been reduced to about one million. Dead birds also showed up in residential areas miles away from where the bait had been put out. All told that winter, about six million starlings were killed.[7]

Emboldened, researchers expanded the DRC-1339 experiments. In Tulare and Fresno Counties, where starlings had been frequenting vineyards and orchards, about twenty thousand were poisoned at two operations. Farther south, in Ventura and Imperial Counties, eight cattle feedlots besieged by starlings were targeted. Often there was so much bait to prepare that cement mixers were brought in to combine the poison with raisins, water, and tallow. This time the bait wasn't placed on specially built wooden stations but sprinkled on the ground between cattle pens. The treatment was deemed mostly a failure, possibly because of where the poisoned bait was used or the warmer, drier conditions in Southern California.

For those trying to control starlings, one of the big upsides of Starlicide was that it was more toxic to starlings than to other birds. But the question became, Exactly how potent did a dose have to be to kill a starling?

In one study at the University of California–Davis, twenty male starlings were given 12.5 milligrams of the poison for every kilogram they weighed. Two of the birds were then killed every two hours—researchers called it "sacrificed"—so their organs could be examined. They showed no signs of organ damage in the first ten hours but generally stopped moving after twelve hours, and most died twelve to twenty hours after they were poisoned. A second, similar experiment tracked the buildup of uric acid in their blood, a sign of kidney function failure. As researchers zeroed in on dosage amounts, they also tried to figure out the best time of day to poison starlings.

In a separate study at the university, researchers experimented on two groups of sixty captive starlings for more than a year, focusing on what time of day the birds were most susceptible to DRC-1339. The starlings were divided into groups and put into light-tight boxes, where researchers then controlled how much light the birds were given. Ultimately, they found that starlings were most likely to die from DRC-1339 if given the poison in the hours around dawn and least susceptible four to seven hours after the fall of darkness. The study didn't answer all their questions—they still didn't know whether the results were due to feeding or other behaviors—but it at least established some ground rules: it was best to get the starlings eating the bait when it was light out, and it probably wasn't worth the bother once night set in.

Still more experiments at UC Davis used flocks of captive starlings to look at whether the birds built up a tolerance to DRC-1339 if given sublethal doses or whether it would kill them faster as they consumed more.

In the winter of 1968 federal scientists and representatives from Ralston-Purina toured several feedlots in California, Iowa, and Idaho to see Starlicide in action. In some cases, it was being mixed with old french fries or oats. Elsewhere, pellets were put out. Hundreds of pounds of the stuff was being deployed with no small effort. The result? Still pretty mixed.

Meanwhile, more studies were carried out that had nothing to do with DRC-1339. In one of them, scientists tried to determine how best to experiment on starlings in a search for ways to inhibit their breeding. They spent a year meticulously measuring the left testes of dozens of captive starlings that been split into two groups, one in an outdoor enclosure and another that was given only six hours of light per day. For the outdoor starlings, the males were in a reproductive state for three to four months, which would be the theoretical window for getting them to ingest infertility drugs.

The six-hours-of-light birds were ready for reproduction for a much longer period of time, making these birds more susceptible to efforts to impede breeding. The study, strange as it was, shed some new light on the starlings' reproductive cycles.

Separately, 357 captive male starlings were enlisted in an experiment testing a chemical that had a sterilizing effect in rodents. Again, researchers spent the better part of the year measuring the size and weight of starling testes in hopes of seeing smaller sizes, indicating a reduced capacity for breeding. The results of the first round were mostly inconclusive. More tests would have to be done.

The frenzy of experiments, hypotheses, hunches, curses, and discussions lasted through the 1960s. A team of bird behaviorists examining where starlings liked to nest went so far as to employ IBM coding forms and punch cards to sort and analyze data. The method established ways to help detect trends that weren't readily apparent from field observations. The initial take was that starlings liked nesting spots that were close to feeding areas, had easy access to temporary roosting sites or staging areas, and were close to water. The most likely reasons to abandon nesting sites included human disturbance, predation from other species, and diminishing water sources.

Elsewhere, radiation was tried. In a series of experiments at UC Davis, about 260 caged starlings were exposed to varying levels of radiation from cobalt-60, a synthetic radioactive isotope. They were then observed for months and examined after death. Those given high doses of radiation died within two days. Lower doses proved fatal in a week or two. Still smaller doses killed about one in five birds. The researchers proposed that a low but consistent source of radiation placed near a bait station at a feedlot might be effective in killing starlings over time without endangering people or other animals.[8]

In the Central Valley, experimenters showed up at vineyards and fig farms with explosives, trying to figure out the best intervals for loud noises to flush away starlings. Carbide cannons were mounted atop twelve-foot ladders and ignited every ten minutes or so. Others produced a bang every few minutes during the day. The noise concussions did the trick most of the time, sending starlings fleeing. Some were gone for the rest of the growing season, while others trickled back in after the ruckus stopped.

Finally, there were some larger, more baffling questions when it came to California's starlings: Where did they come from, and where did they go when they left?

Between 1961 and 1964 more than forty-one thousand starlings were trapped across California and banded. Most were never seen again, but through 1969 about 850 bands (or banded birds) were returned with location information, giving scientists a slightly less muddy understanding of the starlings' flight patterns. Some starlings, of course, stayed close. Those that remained in California turned up, on average, about sixty-two miles from where they were first captured and banded. Other, more adventurous ones traveled five hundred miles or more, reaching the fruit orchards of Washington State, feedlots in Idaho, and even into British Columbia. Their flight patterns tended to follow a roughly diagonal line, moving southwest–northeast, arriving in California in the fall and leaving in the spring. And they seemed to exist in two populations, one in Northern California and the other in Southern California, with the Tehachapi Mountains acting as a fairly firm dividing line. While the northern birds tended to fly deeper into the Pacific Northwest, California's southern starlings often ended up in northern Utah.

The team of researchers on the banding project still couldn't get some of the answers they wanted, like the proportions of the state's resident versus migratory starlings or how often birds that left California actually returned. That would have to come later. One thing that did become clear was that while some starlings stayed in California, many didn't. So even if they could be controlled in the Golden State—still a flummoxing proposition after a decade of experiments—these birds traveling in their mesmerizing flocks and with insatiable appetites recognized no state borders. And all the while, starling numbers in California continued to climb dramatically in the 1970s, great plague-like flocks devouring with abandon, especially the grape crops.[9]

Exasperated, state officials released a biting radio piece about starlings in 1965 to CBS, Armed Forces Radio, Voice of America, and other international outlets, calling them "Public Bird Enemy

Number One." This bird was a marauder, a menace, a criminal, or worse. "Most of us might think killing is too good for this devilish creature," the story said.[10]

California had ultimately failed to discover a silver bullet for this public enemy. If there was any hope of help for the starling problem, it would need to come from somewhere else. Or perhaps *someone* else.

17

Rise of the Bird Men

JAMES A. "JIMMIE" SOULES WAS THE CITY FOOD INSPECTOR in Decatur, Illinois, in the 1940s. As he made the rounds of restaurants and filling stations, it was impossible to miss the starlings wheeling overhead, especially downtown. The first ones had shown up in 1934, and by 1939 they were all over the heart of the city. Soon more than a million were babbling away at a park in the middle of town. When they took to the air, it was like a great screeching cloud lifting off the earth, the noise bouncing off buildings and thundering through the streets.

Predictably, a war on starlings in Decatur ensued. Attorneys with offices near the park stood in the windows and plinked the birds with BB guns and small-caliber handguns. Citizen shotgun squads under police supervision were sent out to fire at the roosting birds. As elsewhere, the fire department sprayed water, tree branches were electrified, and poisoned grain was left out. The biggest gains came in 1946, when ten thousand starlings in an old church tower were gassed with carbon monoxide. Still, giant flocks of birds persisted atop telephone wires, building ledges, and in just as many trees as ever.

Like most everyone in Decatur, Soules, a former game warden, kept close tabs on the city's starling situation. It was as frustrating as it was fascinating. At one point, Joe Swisher, head of the city's pest office, put out some taxidermied owls in the trees. The tactic worked to scare the starlings away, but the stuffed owls didn't last in the midwestern snow and cold. Soules and Swisher decided to join forces on Swisher's owl idea but with an important twist. Working out of a garage in Decatur, their Starling Pest Control Company began producing life-size fake owls made from aluminum, each with luminous yellow eyes made of glass and feathers fashioned from wool and rayon. The owls, available in brown, gray, or white, came mounted on a fake branch. Each had an extra owl face painted on the back, meant to spook starlings approaching from any direction, especially those coming into the city in the evening after a day out in the wilds. "We found that starlings came to the city because they were afraid of owls in the woods," Soules was reported as saying.[1]

Their phony owls were put up around town, sometimes in every third tree, and the frightened starlings were steered toward traps: an old church steeple on one end of town and a set of chicken-wire enclosures on the other. The ones caught in the church, which totaled around five thousand during the first attempt, were quickly gassed. Another ten thousand or so were caught in the wire traps and also killed. "An army of citizens with shotguns used 800 shells and knocked off 6,000 more," the same report said. Decatur's starling problem was starting to go away, and the heroic owls made the news.[2]

Soules and Swisher, well aware of the emerging starling problem around the country, decided to go bigger. They set up a mail-order operation from the garage, selling the owls for $10 to $15 apiece. By the end of 1947 they had produced about three thousand owls, which had been deployed in six cities.

Ten of them went to guard an opulent art deco building in downtown Cincinnati called the Cincinnati Club. The building, like much of the city, was beset with starlings each winter. In the fall of 1947 local residents were already dreading the arrival of birds by the millions. By then it was clear these were descendants of Eugene Schieffelin's starlings in Central Park, not the long-gone ones released by Andrew Erkenbrecher in Cincinnati. Not that it really mattered. "Black automobiles will be white again and there will be much wailing and gnashing of teeth," a local newspaper columnist wrote about the impending annual arrival of the starlings. The columnist had spoken to the club's manager about the fake owls.

"He expects these owls to scare the hell out of the starlings—scare them so badly that they won't want any more of the Cincinnati Club as their happy home for the winter."[3]

Not everyone shared the manager's enthusiasm. The man who ran the Workhouse, a jail-like facility on six acres at the edge of town, had become something of an expert on battling starlings over the years. He'd already tried releasing cats, using poison, and even borrowing live owls from the Cincinnati Zoo. The only thing that really worked, he said, was firepower. He claimed that riflemen had slaughtered three thousand starlings in one night. What good could a few fake owls do when live ones didn't even make a dent in their numbers? "Hell, starlings sleep with 'em," he said.[4]

Despite skepticism, the owls sold. City officials were sick of solutions that never seemed to work and were willing to try anything new. In Lancaster, Pennsylvania, four of the fake owls went up on the country courthouse building. They also showed up in Wichita, Kansas; Louisville, Kentucky; and Muncie, Indiana.

A team from *Life* magazine visited the company's headquarters in Decatur in 1947 and tagged along with Soules and Swisher as they installed owls at the Illinois capitol building in Springfield. Millions of starlings had left behind eleven tons of droppings at the capitol the previous winter and were back for another stay. The problem was so bad that workers sometimes left the building with umbrellas deployed overhead to shield themselves. The fake owls did the trick, *Life* happily reported in a three-page spread, and the starlings left for "less desirable roosts in the city."[5]

Soules and Swisher then told the Springfield city council they could get rid of all the city's 1.2 million starlings for $4,000 to $8,000, depending on the number of owls needed. By the spring of 1948 Soules reported their owls had been sold to customers in forty-four states as well as in Canada and Cuba.

Back home in Decatur, though, the starlings began to return a few years later. The birds had grown accustomed to the decoys, sometimes even standing on the owls' heads in triumph. Soules blamed city officials for not following the instructions on how best to use the owls, but a larger point was being made: the business of dealing with starlings was more complicated than just deploying aluminum owls. It was also clear that the starling problem was here to stay in cities across the United States, and there was money to be made by someone with the right amount of hustle, enterprise, and marketing savvy.

Soules and Swisher eventually parted ways, and Soules set up a new company in Decatur, the Bird Repellent Company. By then Soules had become something of a celebrity—almost certainly the best-known man from Decatur at the moment—and saw an opportunity to expand his empire, including harassing birds other than starlings. In his city, he kept busy chasing birds from the local schools and was hired to get rid of two thousand pigeons roosting at the First Methodist Church. He and his son got to work, a local newspaper said, "with shotguns, traps and skillfully swung tennis racquets."[6]

While the tools of the trade tended to be physical items—fake owls, traps, and the like—Soules was really selling himself as a sort of savant with a supernatural connection with his quarry. "Some people say I even think like a bird," Soules said. "When a pigeon flies, I can tell where he's going."[7]

With starlings, the work was 95 percent psychology, Soules claimed. He declined to disclose the psychological aspects—trade secrets and all—but professed that when tired starlings came home to roost after a busy day of flying and eating, they weren't happy to see Soules or his owls. After repeated hassled homecomings, the starlings simply chose to go somewhere else. Mission accomplished, at least for the buildings he was trying to protect.[8]

He made headlines in 1959 as he and his company spent three months in St. Louis trying to rid nearly a hundred downtown buildings of starlings and pigeons. The crew worked at night mostly, sometimes tracking the birds with spotlights. The $60,000 contract drew scrutiny and intense interest, especially around Soules's methods. Soules took pains to cultivate an air of mystery. He often showed up with a rectangular black box about four feet long but declined to show onlookers what was hidden inside, claiming it contained seven types of proprietary treatments that must remain secret. Sometimes he left the box on a perch popular with starlings and pigeons, as if its mere presence were enough to deter birds. At one point Soules announced plans to publicly disclose the contents of his "magic black box" but then changed his mind. Locals speculated it contained a high-frequency whistle, special lights, or even a gun. "But Soules insists he would never harm any bird and that the box simply 'educates' the birds to go to their natural habitats, the forests and farms, where they might be of some good," a local newspaper said.[9]

While left in the dark about many of his methods, city officials in St. Louis said they were ultimately happy with Soules' work that year in clearing out the birds. He built on that success, expanding his business over the decades to work throughout the country: city halls, state capitols, a nuclear power plant, a pet food mill, a tire factory, banks, and city parks. Eventually he shared some of his methods, noting that he'd experimented with glues, greases, high-pitched sounds, electrified wires, and noisy recordings. While he tried not to kill birds, he admitted to using poisoned bait on occasion and shooting the most recalcitrant. "All of us are crack shots," Soules said about those who worked for his company, including his son and eventually his grandson. "I could knock a nickel off a ledge five stories high."[10]

No matter how many rifles and poisons and fake owls were deployed, there were always more starlings to chase, chambers of commerce begging for help, and frustrated statehouse officials sick and tired of cleaning up bird droppings. So there was room for another miracle peddler for America's peskiest bird problem.

ABOUT TWO HUNDRED MILES NORTH OF SOULES'S OPERATION in Decatur, in the little town of Skokie, Illinois, another starling whisperer came on the scene, a prideful and buttoned-up little man. Joseph Fink, along with his parents and siblings, had come from Lithuania in 1904 and settled on Chicago's Northwest Side, where the family operated a produce stall and sold poultry door-to-door. After graduating from the University of Illinois, Fink started a battery business and eventually built a summer home near Geneva, Illinois. The house was often beset by pigeons and starlings. Fink and a neighbor, who was a chemist, began working on a compound to keep birds off ledges and out of trees. They came up with a gooey formula but couldn't figure out a way to get it to adhere to metal. On vacation in Hawaii in 1947, Fink noticed a plant whose big leaves kept sticking to his trousers. There was something on its leaves, an adhesive, that might prove useful. He cut a sample and packed it between his underwear and pajamas for the flight home. Extracts from the plant wound up being the solution to the problem with their goo, and Roost No More was patented the following year.[11]

The product was sold in an aerosol can about the size of a can of shaving cream. Application was simple: hold down the nozzle and spray in the places where birds had been roosting. The chemical, thick and viscous, gave the birds a "hotfoot," Fink liked to say, but was otherwise harmless. "When the birds step in it, they get itchy feet and don't want to undergo the experience anymore," Fink said. He also claimed that birds communicated with each other when a roost had been treated with the chemical, acting as a preemptive warning that kept them away.[12]

The first application was on an eight-story courthouse in Danville, Illinois. "If those birds are still roosting on your courthouse in three months, you don't have to pay me," Fink told local officials. The goo worked and Fink got paid. Other jobs followed for Fink's company, now known as National Bird Control Laboratories, but his big break came in 1953 in Washington DC.[13]

Every four years, people poured into the nation's capital for the presidential inauguration and parade. The event in January coincided with the starlings' annual visit to Washington, and inevitably, on the big day, thousands of them flocked to the elms and oaks lining Pennsylvania Avenue and other spots along the parade route. "Tenacious as office-holders, persistent as lobbyists, insensitive as social climbers—all familiar types here—the starlings of Washington will not quit," a *New York Times* columnist in DC wrote in the mid-1950s. "Every afternoon at 5, the great hour of social assembly in the capital, the starlings gather in the trees and buildings along Pennsylvania Avenue—a choice of rendezvous that becomes a matter of municipal frenzy, not to say international concern each Inauguration Day. . . . In years past, furious tourists and indignant foreign guests have suffered intolerably from the attentions of the roosting starlings."[14]

The birds had become such a distraction and nuisance that before the first inauguration of Dwight D. Eisenhower, someone decided to do something about it. That's how Joseph Fink found himself in Washington in January 1953, coating as many trees as he could with his gooey Roost No More. It was successful enough that he won the job for Eisenhower's second inauguration and then held the gig for decades. The quirky story of the Bird Man from Skokie, Illinois, was irresistible for reporters, who splashed him across the pages of the *Wall Street Journal*, the *New York Times*, and *Newsweek*.

His fame grew in 1963 when he traveled to England, hoping to sell his services in places like Manchester, Birmingham, Leeds, and Liverpool. In London, he stopped at Trafalgar Square, where several thousand pigeons were gathered. "That place needs my service," he told the Associated Press during the visit for an article that featured a photo of Fink, bald and bespectacled in a suit and tie, with a pigeon perched on his head.[15]

By the fall of that year Roost No More was being used in nineteen countries and had been applied to 150,000 buildings, including the Empire State Building, Fink claimed. As he traveled through Europe, he secured a distribution deal in England and had a party thrown in his honor in Paris at the Hôtel de Ville, where Charles de Gaulle reportedly showed up. All the while, he was constantly being sought after to work his magic with the birds. "I had to check out of the hotel and go to the country just to get a rest," he said.[16]

Over the next decade, Fink's portfolio grew, including clearing starlings, pigeons, and other birds from twenty-two state capitol buildings. Fink later told a story about spending three years trying to get the contract to shoo birds from the capitol grounds in Albany, New York. Luck finally intervened when a pigeon took aim at the Illinois secretary of state, who was visiting Albany and complained to state officials. Fink had the $3,800 contract a week later and was soon collecting $50,000 from the state of New York each year to chase birds around the state. He secured similar jobs when a railroad president was bombed by a pigeon in the Midwest and the daughter of a generous benefactor to a church got similar treatment.

In Washington DC, the high-profile inauguration job paid handsomely too. In 1969 Fink's company received $10,000 for its services, which by then involved a crew of eight men and the treatment of ninety-six trees. By 1973, for the second inauguration of President

Nixon, the price had gone up to $13,000 to spray six hundred gallons of his gunk on the trees. "We want the only white on top of hats of dignitaries to be from the sheen of reflected sunlight," a company spokesman said.[17]

Like Soules, Fink rarely shied away from press, and his reputation got him work in Japan and Israel, as well as a call from government officials in the Soviet Union looking for advice on coping with pesky birds. He bird-proofed Plymouth Rock, Boston's city hall, and a statue of King George V in Brisbane, Australia. "He is too little appreciated," *Sports Illustrated* gushed about Fink at one point.[18]

He may have reached peak exposure on April 6, 1973, when he appeared on the popular TV game show *To Tell the Truth*, where four celebrity panelists grilled him and two impostors to try to figure out who was the real Joseph Fink. Nipsey Russell and Kitty Carlisle were the only ones to guess correctly, and Fink, looking a little impish, couldn't resist an on-air plug for Roost No More. "You can buy an aerosol can for $2.95 at any hardware store," he said.

OTTO STANDKE, THE CIGAR-MUNCHING BIRD MAN INVITED to Mount Vernon, New York, in the late 1950s, also came from the Midwest. And like Soules and Fink, he understood that part of the job was chasing birds and that equally important was making a name for himself in any way possible. But his reputation took a hit after he failed to drive starlings out of Mount Vernon with his paddles, chime, and mysterious gray box in the summer of 1959. The man from Great Bend, Kansas, collected just a few hundred dollars from the city after falling short of his contract, which promised to pay $4,000 if he fulfilled his grand promises of keeping the birds away. "The last of the country slickers," the *New York Daily News* had called him.[19]

Standke, chagrined but not defeated, moved on. Indianapolis was one of his next stops, where he was offered $4,000 to go after the starlings occupying the window ledges of at least six buildings downtown. His efforts worked in some places but not everywhere, and none of the results were permanent. Either way, he always found time for showmanship and self-promotion, perpetually coy about his methods and cocksure about his results. "He spends the days under hot television studio lights, and the nights clambering over the Indiana capital's tallest buildings," a newspaper story said.[20]

He bragged about ridding starlings from buildings as high as seventeen stories with his clandestine techniques. His promotional handouts claimed success at federal buildings in Indiana, Kentucky, and Iowa; two dozen buildings in Topeka and Wichita; and others in Arizona and Ohio. One of his flyers sang out:

WAR! WAR! At Last A War That Will Benefit Everyone—A war on starlings

The Bird Man, Otto D. Standke, can rid any place of the starling nuisance, and they will not return. A proven method.

Can also rid places of pigeons.

He made it a point to go wherever he was called, within limits. "No service on the Moon," his business card quipped.

In the 1960s the *Elks Magazine* ran a three-page feature on Standke called "The Great Starling Chase," which included a big photo of him grinning with a cigar in his mouth and a starling in his hands. The story lionized Standke and played along with his insistence that his methods remain top secret and that all other methods of clearing starlings were bunk: "Otto is to starlings what Saint Patrick was to the snakes of Ireland and what the Pied Piper was to the rats of Hamelin. The Birdman's white-thatched head won't rest easy until every last one of the vexing blackbirds has been banished from our cities."[21]

But he was also hounded by skepticism, even more so than Soules and Fink. "Where comic stops and exterminator begins in Standke is an indefinite line," the *Indianapolis Times* said in 1962.[22]

In 1964 a columnist with the *Post-Dispatch* in St. Louis, where Standke had been hired years earlier to confront the city's starling problem, noted that his star was falling. "Otto belonged to the fading tradition of the pitchman, the rainmaker and the medicine show spieler. He had style and a gimmick—the mysterious dark box, the inside of which no one but Otto ever saw," the article said. "When Otto worked, he was a lot like a third-base coach in baseball.

He constantly gave decoy signals. . . . There is something diverting about a 70-year-old man in a checkered hat and white shoes running around ringing chimes and knocking a pair of pancake turners together." The columnist noted that a friend of his claimed to have gotten a fleeting glimpse into Standke's double-locked box and saw two dead starlings inside.[23]

Standke loved to tease about the contents of the box. "I keep it under my bed at night so no spies will get ahold of it," he once said. Another time, asked about its contents, he said, "Dehydrated blondes. No, I'm just kidding. There's really only an old moldy sandwich."[24]

For all the public interest that he generated, Standke never found his way into the inner circle of serious researchers trying to find a way to deal with starlings, especially the government agents, whose work rarely made the papers. Johnson Neff, who spent decades at the U.S. Fish and Wildlife Service's Denver Wildlife Research Center working on ways to protect agricultural operations from birds, wrote a letter to a colleague in 1965 that blasted Standke not only for his methods but also for being a publicity hound. "He's an old cutie, a gimmicker of top-level stature," Neff said. Once, Neff recalled, a coworker had surreptitiously followed Standke on a job in the Midwest to see how he used his secret instruments. Little was revealed, and Neff speculated that any success Standke had was likely due to simple harassment of starlings. The birds probably flew away as Standke crawled over building ledges and rooftops and made a nuisance of himself in the trees. "The old man is nothing if not persistent. He works at the job and works hard."[25]

Neff remained deeply suspicious of any long-term gains that Standke claimed in keeping birds away and was frustrated that the U.S. General Services Administration—the federal agency that acts as a landlord for federal buildings—would pay Standke money when government researchers were developing their own methods for dealing with starlings. But the sourest feelings seem to have been directed toward Standke's outlandish promotional behavior and that damned box he never seemed to stop talking about. "The question that has been most often asked of me is: What does he have in the locked box? Nobody, I mean nobody, knows,"

Neff said. "Personally, I do not believe that there is anything in that box that is germane or pertinent to starling frightening."[26]

Speculation about the box continued even after Standke died in 1970. One building manager who had contracted with Standke believed it contained a rope made of some kind of metal that the Bird Man used to flail at the starlings. Others thought it held a noisemaker or a special chemical or perhaps nothing at all. The *Indianapolis Star* did some digging in 1973 and found that Standke, before he died, gave his special box to his landlady of thirty years, Virginia Andress. "I knew all about it," she told the *Star*, "but I'm not telling because I know he didn't want anybody to know." Great Bend attorney Arthur Hagen said he had helped handle Standke's estate and took a peek inside in the box. "There was some poisoned wheat which, I think, was used for pigeons. Then he had some gadgets to make some noise with," he said.[27]

As Neff pointed out, it's possible that the secret to Standke's success, when there was success, was simply his ability to spend hours and hours chasing starlings around city buildings and streets, doing whatever he did to get them to fly away, however briefly. The box, even if just for show, helped prop up the feeling that the mystery of the stubborn infestations of starlings required a solution just as mysterious.

The *St. Louis Post-Dispatch* columnist summed it up: "The world . . . has been ruled by superstition, mystery and strange black boxes a lot longer than it has by science, and there is a certain sadness in the passing of the itinerant bird chasers—the men little boys would flock to the town square to watch—doing their work with arm-waving, incantations and suggestions of dark necromancy. Still, science must be served and birds must be driven away for good. Banks must be kept clean and statues unsoiled."[28]

18

Death from Above

THE SUN WAS SINKING LOW IN THE SKY EARLY IN THE EVEning of October 4, 1960, when Eastern Airlines flight 375, headed for Philadelphia, made its way to runway 9 at Boston's Logan International Airport, revving its engines in preparation for takeoff. It would be dark soon, as the sky traded the sun for a stunning full harvest moon.

On board flight 375 were five crew members and sixty-seven passengers, including fifteen new U.S. marines on their way to training at Parris Island, South Carolina; businessmen heading home after a shoe convention in Boston; baseball fans heading to Pittsburgh to watch the Yankees take on the Pirates; a grandmother on the way to see her grandson; and an engineer carrying a locked suitcase holding plans for a missile defense system. The plane also carried twenty-five hundred pounds of U.S. mail and plenty of fuel for the three-hundred-mile flight to Philadelphia. From there it was scheduled for Charlotte, North Carolina; Greenville, South Carolina; and finally, Atlanta.

About seven seconds after takeoff at Logan, the plane's numberone turbo-propeller engine on the left side seemed to choke out, so the crew shut it down. Next to it, the number-two engine then lost thrust, and a passenger saw fire shoot out from somewhere on the left wing. On the right side, someone saw a dark smudge near the wing, and propeller engine number four went out. The plane slowed dangerously and began dipping to the left. Desperation set in. The number two and four engines were restarted, but it was too late. The plane veered harder to the left, then seemed to recover momentarily, only to slip again. But then the left wing dipped, the nose pitched up, and the entire plane rolled. Screams of terror filled the cabin as the plane, suddenly in the grips of horrifying gravity, fell helplessly back to earth. About forty-seven seconds after takeoff, flight 375 plunged nose-first into Boston Harbor, breaking into several pieces, scattering people, body parts, wreckage, luggage, and mail into the water. The plane's main fuselage, one witness later said, looked as if it had been "ripped open with a can opener." Fifty-nine passengers were killed, along with three members of the crew, in the deadliest air crash in New England history.[1]

The rescue and recovery lasted through the night and for days and weeks as crews pulled everything they could from the water, the muck, and the slippery eel grass on the harbor's bottom. Then came the painstaking job of reassembling the airplane in a warehouse and trying to figure out why flight 375 had taken a nosedive into the water. The initial speculation centered on the make of the airplane, a Lockheed Electra. The Boston crash was the fifth accident involving an Electra within twenty months. More than 220 people had been killed in those terrifying wrecks, which included wings ripped off in midflight. Just three weeks before the disaster at Logan, an Electra had crashed, flipped, and burned while trying to land at New York's LaGuardia Airport. In March 1960 federal aviation officials required reduced speeds for the 130 or so Electras operating in the United States, hoping it might stem the tide of so many aviation tragedies.

Suspicion in the case of flight 375, though, quickly turned to another—smaller, darker, and wilder—culprit. Two days after the crash, with the investigation still in the very early stages, the morning edition of the *Boston Globe* blasted a banner headline on its front page: "Blame Starlings for Crash." Hundreds of dead starlings had been found scattered across runway 9 shortly after the accident. The *Globe* ran a picture of a mangled one below the headline, explaining the emerging theory that the birds had clogged the plane's air intake.

Soon came estimates that flight 375 had plowed into a cloud of ten thousand to twenty thousand starlings shortly after takeoff. It wasn't out of the question. Starlings and gulls had been a problem at Logan Airport for years, and regular patrols were sent out to shoo the birds away. Sometimes they resorted to shotguns.

Flight 375's engines were sent to their manufacturer, General Motors' Allison Division in Indianapolis. There investigators from the Civil Aeronautics Board took them apart and examined each in microscopic detail. Bits of feathers and bird flesh were found in the engines, along with crabs, which they assumed had been feeding on the dead starlings after the crash. Meanwhile, talk intensified over the dangers that the birds posed at the airport. "Any airplane encountering a massive flock of birds at low altitude, when its air speed is critical, is going to have trouble," Elwood R. Quesada, head of the Federal Aviation Administration, said a few weeks after the crash.[2]

Bird strikes have been an issue since the earliest days of aviation. Orville Wright hit one, possibly a red-winged blackbird, in his Wright Flyer above a cornfield near Dayton, Ohio, in 1905. The first fatality came in 1912, when a Wright Model B piloted by Cal Rodgers, the first man to fly across the continental United States, plowed into a flock of gulls and crashed into the ocean near Long Beach, California. From 1912 to 1959, though, bird strikes were responsible for only two more human fatalities and three civilian aircrafts destroyed. The low numbers were partially because the planes at the time were noisy and fairly slow, giving birds plenty of notice to steer clear. The number was higher for military planes, especially in the 1940s, with the onset of faster jets.[3]

But the 1960 crash at Logan airport, still the deadliest bird strike in U.S. history, was a wake-up call, and a turning point, for the aviation industry. The Civil Aeronautics Board spent two years investigating the crash of flight 375 in Boston. While much of the work was a meticulous reconstruction of the airplane, significant focus fell on the starlings and how they could inflict so much damage on a hundred-foot airplane. Necropsies conducted on some of the dead birds on runway 9 determined they had indeed died the day of the crash. An extraordinary scientist named Roxie Laybourne, who was working at the Smithsonian at the time, was brought in to examine the dark feathers found in some of the engines. She confirmed they were from starlings, and her work on the investigation helped pioneer the use of forensic ornithology—using microscopic techniques to identify feathers and species—which became a regular part of future investigations around aviation bird strikes.

And then came the experiments to determine what happened when turboprop engines ingested starlings. In Indianapolis, GM investigators ran a series of tests using air guns to shoot anesthetized starlings, grackles, and blackbirds into mounted engines of the type used on the Electras. The birds went in two at a time, four at a time, and then eight at a time—all in search of a critical threshold. They got what they wanted in the Allison experiments. Even though these were simulations and not in-flight tests, it became clear that just a few birds could interrupt power to the engines and cause flames to shoot out. More tests were conducted at a wind tunnel operated by Lockheed in Burbank, California. There starlings were tossed into an engine running at different power settings and with variable wind speeds. The engine was able to continue operating after ingesting a few starlings; even if the power was sometimes interrupted, the engine could be restarted. It became more difficult when four to six starlings were sucked in. When eight starlings went into the engine, power was often lost, temperatures rose, flames occasionally erupted, and restarting the engine became "improbable."[4]

Ultimately, federal investigators concluded that flight 375 had flown through a flock of starlings, mostly on the left side of the plane. About seven seconds after liftoff, at least four birds were sucked into engine number one, and about six birds got into engine number two, both on the left side. On the right side of the plane, number three didn't ingest any starlings, while number four took in a small number. Power was lost in three engines, but two of them recovered. Then came a "unique and critical sequence" of events. The near-complete loss of power on the left side of the plane caused it to veer left, and when engine number two flamed out and number four continued to limp along in a weakened state, there was no power on the left but substantial power on the right. The plane dipped, rolled, stalled, and simply ran of time to recover before slamming into the harbor. As part of the investigation, the Civil Aeronautics Board ran a series of tests with a flight simulator where expert Electra pilots were tested in conditions that tried to mimic what happened in Boston to see if they could regain control of the doomed plane as it lost speed and fell through the sky. The situation was so complex and harrowing, investigators said, "that human capabilities of perception, recognition, analysis, and reaction were insufficient . . . to accomplish restoration of positive performance control."[5]

Only a few weeks after the crash of flight 375, another Eastern Airlines flight, this one also an Electra, ran into a flock of starlings at Logan Airport just before takeoff, and the flight had to be aborted. Captain W. H. Jenkins told investigators, "We suddenly became cognizant of a large flock or cloud of starlings. These you cannot see from the end of the runway. Or we could not see them before we started out. They were just suddenly there. Starlings are very excitable birds: they wheel and turn in a flock so that one minute, when you are looking at them, you see nothing, and the next minute, as the cloud shifts, there is a big black cloud in front of you." He continued: "They hit the front of the airplane, resembling machine gun fire. Just brrrrrummm! There wasn't a square inch of my windshield that wasn't splattered with bird remains. I couldn't see a thing."[6]

One of the most intriguing questions after the crash at Boston and the aborted flight was whether the engines of the Electra actually attracted starlings, setting up a dangerous clash between birds and machines. The idea was floated by John Swearingen, chief of the Protection and Survival Branch of the Civil Aeromedical Research Institute in Oklahoma City. Swearingen said that the high-pitched twitter coming from Electras while taxiing was "strikingly similar" to the chirping sound of a field of crickets, a common food for starlings. "This is particularly so when one drives along a country road in the late summer at about 20–30 mph and experiences the fusion of individual cricket sounds into a sustained chorus," said Swearingen, who must have had keen hearing. "Starlings have been observed to cruise at 20–30 mph, and would experience a similar sustained exposure to the cricket sound sources."[7]

That speculation, paired with observations that starlings seemed to be activated around Electras, set off a study at the Oklahoma City airport that started in December 1960 and went into the following year. Teams of investigators were sent to the airfield with binoculars and portable tape recorders to observe bird patterns during takeoff and landings of different kinds of aircraft, including Electras. "When the Electra was about halfway down the taxiway from the terminal, the sound became more noticeable. We noticed a relatively large flock of starlings rise up from the side of the runway," one observer said, noting that the birds seemed to stay to the rear of the plane. "Birds seemed confused; flocks dipped and dived in various directions but maintained their same relative position with reference to the aircraft. During takeoff, the birds followed the plane a short way and then settled back along the side of the runway."[8]

The crickets and the Electra's audible chirp also produced similar-looking high-frequency soundwaves, investigators said, making it more plausible that the starlings sometimes mistook one for the other, drawing close to disaster when they thought they were actually coming in for a meal. Swearingen and his team recommended that Lockheed and Allison, the engine manufacturer, consider structural changes to the Electra's engines or that the planes taxi at a different RPM on the tarmac.

The idea proved controversial, though. Lockheed and Allison did their own studies and found that the sounds weren't similar. The federal Denver Wildlife Research Center, still one of the hubs of starling research, also weighed in a few years later, agreeing with the companies that the chirps weren't all that comparable and raising questions about whether starlings actually relied heavily on sound when hunting crickets.[9]

While the debate over the chirps went on, federal aviation officials decided that it was time to rethink how certain airplane engines were designed. In the summer of 1968 the Federal Aviation Administration issued an advisory about "foreign object ingestion," specifically addressing how to mitigate the dangers of birds in the engines. Regulatory changes were issued in 1974. These specified mechanical modifications to turbine engines to minimize risk from birds and changes to how tests should be carried out to best reflect real-world conditions. More refinements came in subsequent years, but the gist was to manufacture engines hardy enough to ingest a lot of small birds (like starlings) or one or two big ones (like geese) and still be capable of operating.

Civilian aircraft weren't the only ones in danger from starlings and other birds. The U.S. Air Force reported more than fifteen hundred bird strikes between 1956 and 1966, causing an estimated $10 million in damage to military jets and equipment. Gulls, starlings, and blackbirds were the most likely to be involved, and about half of the collisions happened near the ground, below two thousand feet. The presence of water or a migration path near an airport often increased the risks of trouble.

Injuries and fatalities were rare, although a sandhill crane killed a T-37 pilot when it slammed through the windshield in a predawn flight in 1966. But the potential for disaster was always present. "The jet engine is prone to ingest anything movable because of its powerful suction. In fact, several humans have been sucked into these engines during ground operations. The compressor blades are, however, very fragile. When one is broken off it produces a chain reaction of foreign object damage. Thus, the ingestion of even one bird causing the failure of one blade can produce total engine failure," an air force researcher said in a 1967 article on bird strikes.[10]

A sense of particular alarm arose in the winter of 1965–66 at Moody Air Force Base in southern Georgia. Some four million starlings and blackbirds were camping in a wooded bog just east of an airstrip. There had been at least seventeen bird strikes over three months, including one that almost destroyed an airplane. In response—or perhaps retaliation—military crews set off a flash bomb after dark one night, flooding the area with a terrifying five-hundred-million-candlepower flash that shook buildings and stunned people and wildlife alike. "We don't know what we've done," an air force major said that night. "This is the first time this has been tried anywhere."[11]

They vowed to try it a few more times if it showed any promise in clearing away the birds. But the dangerous collisions between birds and airplanes—thrust into the public consciousness with the starling-induced Electra crash in Boston Harbor—were only getting started.

SEVERAL FACTORS HAVE CONSPIRED TO ACCELERATE BIRD strikes in the 1960s and beyond, including faster and quieter jets. The United States has also seen a dramatic spike in the number of commercial flights (from fourteen million takeoffs and landings in 1975 to twenty-five million in 2010) and an increase in some avian populations, including larger birds like Canada geese, white pelicans, vultures, bald eagles, and gulls. As a result, 160 civilian aircraft were destroyed or seriously damaged in wildlife collisions between 1960 and 2010, most famously the "Miracle on the Hudson" flight in 2009, when a US Airways plane went into the Hudson River near Manhattan after hitting a flight of geese. One 2019 study found that bird strikes cost the commercial aviation industry at least $187 million each year.[12]

In the early 1990s federal agencies and scientists launched an intensive effort to understand what was going on. Although many of the incidents involved larger birds, like gulls and geese, starlings and other smaller species were still a problem. A 2003 study estimated that starlings and blackbirds, which often intermingled, were responsible for more than $1 million in damage to aircraft every year, with incidents spread across forty-six states. Starlings alone were involved in an average of seventy-one bird strikes each year in the United States between 1990 and 2001. Damage was typically not serious, but the potential for catastrophe remained both in the United States and abroad. In July 1996 a Belgian Air Force cargo plane carrying a military band flew into a flock of several hundred starlings and lapwings at Eindhoven Air Base in the Netherlands. The plane crashed and caught fire. Of the thirty-seven passengers, thirty-four were killed. In 2008 a Ryanair jet crash-landed in Rome after colliding with a group of starlings as it approached the runway. No one was killed, but the plane was badly damaged.[13]

As analysts have crunched the numbers, they've found that the vast majority of bird strikes happen as airplanes take off and land, often in the first several hundred feet above the ground. So there's now a fleet of professionals, many of them wildlife biologists, focused solely on keeping birds and airplanes apart. Most of the action happens around the areas near airports where birds like to congregate: wetlands, grassy spots, waterways, and tree stands,.

Some of the bluntest tactics remain in use. After hundreds of incidents with laughing gulls at John F. Kennedy International Airport in New York, government agents spent the summer of 1991 picking them off with 12-gauge shotguns. Working five days a week, they killed 14,886 in just over two months. Poison is also deployed, including the starling killer DRC-1339, which is often sprinkled on french fries.[14]

But the work around airports has become more sophisticated and coordinated in recent years. There's now a national tracking system for bird strikes, and protocols are in place for submitting the feathers of dead birds to definitively identify what kinds have been hit. National and international conferences are held about bird strikes, and a cottage industry of contractors, vendors, and inventors has arrived on the scene. Meanwhile, biologists with the U.S. Department of Agriculture's Wildlife Services program work at hundreds of airports around the country to help local airport officials figure out the best way to handle their bird problems. Sometimes they conduct "habitat modifications" to make an airport less bird-friendly, including trimming trees, draining standing water, and even putting up man-made obstacles in flat areas to obscure the sightline for birds, which can't see approaching predators and thus become nervous. Crews also rely on familiar methods to move birds along, like superloud cannons, lasers, or bird distress calls blared at high volume.

The solutions are becoming more high-tech. One company has invented a flying drone that resembles a peregrine falcon, which can be deployed to scare off flocks of birds. Early warning technology is being developed to spot potentially problematic birds and even predict their behavior via algorithms based on ornithological insights.

When it comes to starlings, though, no one's cracked the code on how to force them to renounce their stubborn ways around airfields. One morning in October 2019 a flock of several thousand starlings landed at the Asheville Regional Airport in North Carolina. Fearing for passengers' safety, airport officials shut down the airport, grounded flights, and did their best to scare off the starlings, using sirens, water cannons from fire trucks, and pyrotechnic noisemakers. Drivers even chased the birds around the runway, only to watch them land elsewhere on the airport grounds. The starlings took off a few hours later—maybe because of the harassment, maybe because they grew bored—and flights resumed.[15]

19

Can't Beat 'Em? Eat 'Em

SHOTGUN BLASTS BROKE THE TWILIGHT AIR AT THE CAPITOL grounds in Washington DC in March 1934. Pulling the trigger late on a Wednesday afternoon was South Trimble, a former turkey farmer and congressman from Kentucky who was currently the clerk of the U.S. House of Representatives. He'd aimed his gun at the starlings perched in the trees overhead, members of a giant flock that swarmed the grounds each year. A Capitol policeman quickly approached. "It is my duty to tell you that it's against the law to shoot the birds in the Capitol grounds," the officer said. "Well," Trimble said, "you have done your duty."[1]

Moments later, he peppered the branches with more buckshot, and about fifty dead starlings dropped to the ground. Trimble gathered them up and delivered them to a congressional chef, asking that they be baked into pies. The clerk hosted a lunch the next day with the House majority leader and other top politicians, telling them that the pies were filled with local marsh birds. The diners didn't learn of Trimble's deception until it showed up in the newspapers the next day. Still, the dish got rave reviews.

"I want to tell you, starling pie is good," Rep. Jo Byrns said. "The meat is tender and sweet; so sweet in fact, I even picked the bones. But I want to tell you one thing. I thought I had eaten reed birds and had told Trimble the pie was the best reed-bird pie I had ever eaten."[2]

A few blocks away, perhaps inspired by Trimble's escapade at the Capitol, a DC resident took culinary matters into his own hands. "I put out a spring rat trap baited and got nine starlings in three hours," claimed a man named W. H. Turney. "I cleaned them, put them in salt water overnight and then parboiled and fried them. When I catch my next mess of starlings, I expect to try them baked."[3]

Around the same time as Trimble's starling luncheon, newspapers were running a recipe for starling pie that was "fit for a king": "Make a deep crust, put in a bit of onion, cabbage, tomatoes, green peas and salt pork, then add a couple of starlings well cut up, and bake."[4]

Back then, most knew the old English nursery rhyme with this opening stanza:

Sing a song of sixpence,
A pocket full of rye.
Four and twenty blackbirds
Baked in a pie.

Starlings aren't blackbirds, but since the earliest days of their proliferation in North America, there's been talk about whether a solution might be found in human appetites for the little bird. A few years ago, when the conversation turned to my interest in *Sturnus vulgaris*, a birder friend of mine immediately blurted, "Please tell me they've found a recipe for eating starlings." Recipes are one thing, but convincing people to take that first bite is another.

"The rather strong gamy flavor of the starling's flesh will probably limit its popularity, from a culinary stand point," a government biologist said early on, speaking from his own experience. "When the breasts of these birds have been soaked in soda-salt solution for twelve hours and then parboiled in water which is afterwards discarded, they may be used in a meat pie that compares fairly well with one made of blackbirds or English sparrows."[5]

FAMED CONSERVATIONIST JACK MINER SAW PLENTY OF STAR-lings in the early 1920s in Pennsylvania and New England, but the true shock didn't hit until 1924, when he saw his first starling in Canada at his sprawling bird refuge in Kingsville, Ontario, across

Lake Erie from Cleveland and not far from Detroit. At first there were just a few, "and to tell the whole truth I rather welcomed them," he said. But those few multiplied rapidly. Within five years they had driven away five thousand to ten thousand purple martins that summered at the refuge. They also pushed out red-headed woodpeckers and mourning doves. The starlings were so numerous that their droppings alone killed two thousand to three thousand pine trees on the property, Miner reported. The result was, suddenly, a deathly kind of forest. "It gives one about the same sensation we get when looking through the undertaker's show rooms," Miner said.[6]

Fed up, "we declared war on them" in July 1931, he said. Workers at the refuge hastily built a starling net, "and by September first," Miner reported, "we had caught, drowned and buried over 17,000; but bless your life there were a million came to their funeral." Soon neighbors from Windsor, Ontario, arrived at the refuge to help quell the invasion. Miner figured they caught and smothered about three hundred thousand more starlings. The question soon became what to do with all these dead birds.[7]

"These starlings are eatable," Miner said. "My sister scalded and cleaned 24 in less than an hour. . . . She then made a real English blackbird pie of them. . . . The fact is, if properly cooked, they are a treat on the table and that is one point of hope of controlling them."[8]

The idea hit: Why not truck these thousands of birds back to the city and cook them up? Ultimately, that's what they did, and the birds were given to people in need of food. Miner revisited the idea several years after exporting his dead starlings to feed the hungry. At a meeting of the American Wildlife Institute in Detroit, Miner and others formed, only half-jokingly, the Starling Luncheon Club. The group included journalists, conservationists, and others frustrated with starlings, including political cartoonist J. N. "Ding" Darling, president of the National Wildlife Federation and former chief of the U.S. Biological Survey. "The organization's purpose is to control the starling by popularizing it as a food," the group said.[9]

The lunch club never took off, but the idea has kicked around for decades: Could we ever eat our way out of the starling mess?

Starling consumption has gone in and out of vogue for years, at least on a small scale where such a thing was even known and considered. Of course, wild game birds were once a popular staple for food, their attraction ultimately diminished by the convenience of domesticated chickens and then the rise of vast industrial operations cranking out perfectly packaged meals for mass consumption. But putting a starling on your plate? Was it exotic, practical, or merely unseemly?

American mystery writer Rex Stout often had his hero detective Nero Wolfe dining each spring on a platter of starlings. Stout told *Sports Illustrated* in 1956 that April is the best time for a plate of starlings. "Mr. Stout allows four birds to a guest and may shoot a few more than necessary as insurance against stringy oldsters or those hopelessly impregnated with shot," the magazine said. Once the feathers were plucked, he'd marinate the birds in red wine for twelve hours and then broil them. Over moderate heat, they'd be done in twenty-five to forty minutes, after which he'd add a sauce, preferably an herb-based béarnaise with fresh tarragon. "Flavor to taste," Stout noted, "and deliberate a bit over whether or not half a bay leaf will add just about the right touch."[10]

In the early 1960s Herbert Deignan, ornithologist and bird curator at the Smithsonian in Washington DC, took up the idea, claiming that starlings not only were a nuisance but also were "good eating" and faced no legal limitation on their killing. The only problem, he said, was one of preparation and marketing. He recommended promoting starling pies with their feet sticking up through the crust, not unlike the way passenger pigeons had been served at Delmonico's in New York City nearly a century before. "Now if we could only work up a market for them," he said. "We'd have to give them a fancy name, *merle de Maryland* or Virginia quail or something. We could net them the way they do in Italy and if we could market them, I think we could control them."[11]

In the fall of 1966 the Canadian news magazine *Maclean's* ran a story headlined, "Why Not Try a Delicious Starling Pie?" Starlings had shown up in particularly copious numbers in Vancouver, British Columbia, where authorities estimated there were 400,000 roosting under a single rickety bridge and another 1.6 million scattered about the city. The birds were fond of eating breakfast, lunch, and dinner in the nearby Fraser Valley—to the tune of about $200,000 worth of fruits and vegetables in 1965—before settling down for the night in Vancouver. "So far the starlings have cackled in the face of every weapon brought to bear against them," the article said. "There remains one last hope for wiping out starlings: Human gluttony." Indeed, the *Vancouver Sun* had commissioned its food editor to bake a starling pie, following a recipe from an English cookbook from 1898. Staffers sampled the dish. "Great," one of the editors said. "They taste something like Scottish grouse."[12]

For a few months in 1971 a debate quietly raged at the U.S. Fish and Wildlife Service's Denver Wildlife Research Center about whether promoting the hunting and eating of starlings might be a worthwhile tactic for reducing the number of starlings and, in the process, helping native birds that were being harassed and displaced. "I have hunted and eaten starlings and find them quite satisfactory as table fare," said biologist Earl Baysinger in an interoffice memo, kicking off a series of spirited communiqués on the subject. The federal government could kick-start the initiative, he said, by classifying starlings as huntable game birds and perhaps providing some recipes for the quarry. "We already have the birds," Baysinger said. "All we need is to introduce hunters to them." He and others noted that there'd already been some recent dabbling in the practice. In Idaho, starlings had been served at several dinners for the Rotary and Kiwanis Clubs. And in the summer of 1971 a hundred starlings were given to the Yakima Audubon Club in Washington State. An Audubon representative reported, "Most of the members agreed that the meat was comparable to dove or pigeons."[13]

The back-and-forth at the Denver Research Center included a 1965 Associated Press story from Portugal about how starlings and sparrows had become a delicacy "to be savored with wine" in Lisbon. The birds were typically caught in nets after being baited with corn or by glue applied to tree branches that trapped starlings in place long enough for local boys to climb up and retrieve them. The birds were fried until crunchy and consumed by the thousands at restaurants in the city. Could it catch on in America? "I doubt if the American people have a craving for starling meat," one of the researchers wrote.[14]

Another federal wildlife official noted that when both red-winged blackbirds and starlings were killed and left on the ground, dogs and cats "always preferred" the blackbirds. Someone included a newspaper clipping about a daring state biologist in Indiana who put his body on the line in the name of research. "He prepared starlings for the table in every way imaginable—boiled, fried, marinated, broiled, stewed, baked, casseroled, sauteed, and ground for starling burgers," the story said. "Following this dreadful assault on his stomach, he summarized his experiment in one sentence: 'Horrible, tastes like liver that's been left on the sidewalk for a week.'"[15]

Even if people developed a taste for starlings, could the general public be trusted to identify and shoot the right birds? "Remember when they killed 3,000 martins in Jefferson City MO and thought they were 'starlings.' Most people can't tell the difference between starlings and blackbirds," someone scribbled in the margins of one of the memos. "How about hiring and training some people on welfare to hunt and trap them, then give the meat to poor people who want it?" another member of the team suggested.[16]

But nothing seemed to gain traction among a majority at the research center. "I don't believe sport shooting of starlings would ever catch on in the U.S. in a big way. Even if I were wrong on that score, I don't think shooting could make any real dent in the continental starling population," one of the final notes read. "I don't think starling meat will be widely accepted and consumed as long as we have an ample supply of beef, poultry, etc." The writer concluded, "I think it has no real merit from any standpoint."[17]

Still, the idea never quite vanished. In 1992 a top starling researcher with the U.S. Department of Agriculture gave the keynote address at a symposium on regional food and wines in Ohio. He delivered a detailed history about birds in Ohio, particularly how passenger pigeons had been replaced by starlings and blackbirds after the great forests were cleared. When starlings moved in and discovered crops growing in the state's rich soil, inevitable conflicts arose, and people attempted all manner of ways to kill the birds or rid the landscape of them. The researcher, though, offered a two-pronged path toward coexistence: a little more tolerance and a tad more appetite. "So let us manage Ohio's bounty of blackbirds and starlings not merely as pests that are slaughtered and left to rot—but rather as a renewable resource that is grain and grape fed, delicious, nutritious, and USDA approved, Grade A Prime. Bon appetite [sic]!"[18]

IN 2004 AN ARTICLE BY CONSERVATION BIOLOGIST AND WRITER Joe Roman titled "Eat the Invasives!" was published in *Audubon* magazine. The story called for readers to enlist their mouths and stomachs in the fight against invasive species like wild boars, nutria, carp, and kudzu. At the time, he estimated that invasive animals and plants had taken over a hundred million acres in the United States, endangering native species, threatening water and soil, and costing about $138 billion each year. More often than not, eradication of nonnative species had a failing record, even after the use of poisons, fire, biological controls, and all-out war. Perhaps we've overlooked one of the most powerful weapons, Roman suggested. "There's already a relentless omnivore on the plains, in the mountains, along the shore, and perhaps in its favorite reading chair right now," he wrote. "In fact nothing is better qualified to wipe out an entire species than humans." After all, hadn't we already wiped out billions of passenger pigeons and millions of bison, not to mention our decimation of Atlantic cod and right whales? "Why not turn this enthusiasm on the very invaders we're trying to fight?"[19]

He offered several recipes, like egg rolls stuffed with nutria and wild boars, carp served in sour cream with wild mushrooms, and "Exotic Jade Soup" featuring invasive weeds like dandelions. Afterward, his Eat the Invaders group ("Fighting invasive species, one bite at a time") expanded to include recipes for dozens of other invasive species like watercress, garlic mustard, periwinkle snails, Asian shore crabs, and sow thistle. The "invasivorism movement," as some called it, had found some momentum.[20]

An invasive species banquet was hosted by a conservation group at the James Beard House in New York City in 2011. The National Oceanic and Atmospheric Administration in 2015 launched its "Eat Lionfish" campaign to employ human appetites against the home-aquarium fish that was threatening coral reef species in the Southeast and the Caribbean. "Once stripped of its venomous spines, cleaned, and filleted like any other fish, the lionfish becomes delectable seafood fare," the government agency said, adding, "If we can't beat them, let's eat them!"[21]

Every year since 2012, except during COVID, the Institute for Applied Ecology in Corvallis, Oregon, has hosted the Invasive Species Cook-Off, also known as Eradication by Mastication. Featured items include blackberries, clams, bullfrogs, dandelions, and thistle. Starlings show up occasionally. One contest featured starling bacon kebabs with a blackberry balsamic reduction. Another included a pâté made of opossum and starlings. In 2016 the winning entry in the "savory meat" category was a plate of blackberry lime starling tacos.

For all its novelty, advocates remained realistic about the effects of a gastronomic approach to unwanted species. "We don't really think we're going to eat our way out of this problem," said Tom Kaye, the institute's director, during the year of the starling tacos. "It's really an awareness step, to bring people to the table, literally, to eat them and talk about them and the damage they're doing in a more light-hearted environment because, let's face it, some of the impacts they're having are pretty devastating and kind of depressing. It's important to maintain our enthusiasm in the face of a problem like that."[22]

20

Mapping the Travelers

IN THE SPRING OF 1970 A STARLING FLEW INSIDE THE CHIMney of a house just outside the Canadian village of Linden, Alberta, about sixty miles northeast of Calgary. A hood at the top of the chimney prevented the bird from escaping, so it ended up traveling down a vent pipe and popping out in the basement. The homeowners killed the wayward bird—exact methods weren't disclosed—and soon found a tiny transmitter, weighing less than a gram, attached to its tail feathers with the number 732-49729. The device, along with the little metal band on its leg, belonged to the U.S. government.

The starling had been fitted with a transmitter as part of an experiment that began outside Portland, Oregon, where a flock of about 180,000 had descended on a holly orchard, fouling the leaves and killing many of the trees. In February a dozen starlings at the farm were caught in a net, weighed, measured, equipped with a radio transmitter, and released. Scientists with the federal Denver Wildlife Research Center used antennas to track the birds' movements, picking up periodic beeps that gave away their location. For the first three nights number 732-49729 returned to the orchard. Several nights later it showed up at another holly orchard about nine miles away. And then it disappeared, and scientists wrote it off as another lost bird. They were stunned to get word more than a year later that it had been found in Canada. Number 732-49729 had flown more than eight hundred miles, finishing its extraordinary journey two months after its capture in a basement a considerable distance from others in its flock. "It is by far the greatest movement reported for a starling with a transmitter attached," biologist Olin E. Bray, who helped pioneer radiotelemetry tracking of starlings, grackles, and other "problem" birds, said at the time.[1]

A few years later the flock at the Oregon holly orchard had grown to more than 800,000. Bray and his team ran another experiment, capturing eight starlings and clipping transmitters on their tails. The team tracked the birds for fifty-one days, carefully observing them as they moved among the orchard, pastures, grainfields, waterways, and tree roosts. On average, they had a home range of around twenty-four miles. None of them took off for Canada or replicated the epic flight of number 732-49729. Eventually, the homeowner in Linden mailed the transmitter to Bray but asked if he might get it back someday. He hoped to put it on display at a local school, along with the story of the bird and its amazing trek to this little town.

PART OF WHAT'S SO OFTEN INSCRUTABLE ABOUT BIRDS IS that despite their proximity to us, we feel as though we know little about their daily lives, especially their extraordinary travels. "On their journeys, the migrants not only travel vast distances overland but also cross pathless seas and oceans," British ornithologist William Eagle Clarke said in 1912. "The question is—how do they find their way? How are they guided? Here we are face to face with one of the greatest mysteries to be found in the animal kingdom."[2]

Birds don't stay put for a lot of reasons, but the main two are their needs for eating and breeding. Birds that nest in the Northern Hemisphere typically fly north in the spring to feast on emerging insect populations and plants in bud and to find nesting sites. They head back south with the onset of winter as insect populations decrease and food becomes scarcer. There are exceptions and variations, of course, since nature always seems to insist on a few.

More than half of the bird species in North America are migratory. Some move very short distances, such as higher or lower on a mountain. Others may fly a few hundred miles. Still more are long-distance migrators, journeying thousands of miles each year. Red knots, for instance, move annually from their wintering grounds in Tierra del Fuego, Argentina, to their Arctic summer breeding grounds in Canada, stopping along the way on the Eastern Seaboard of the United States.

The triggers of migration can be complex and subtle, a mix of changes in food availability, temperature, daylight hours, and some internal signal, well honed by eons of evolution, that says it's time to move along. For some species, smell, hearing, and the ability to key in to the earth's magnetic field may also factor in. The exact mechanisms aren't fully understood. Complicating things, not all birds within a species migrate in quite the same way or even migrate at all.

But those birds that do migrate require a complex combination of orientation and navigation, the sense to know where they are, where they need to be, what season it is, and what the diurnal cycle is. And finally, they must possess the ability to get from point A to point B without running into major trouble. "In short," ornithologist Ian Newton wrote in *The Migration Ecology of Birds*, "they need the equivalents of a map, compass, calendar and clock, together with a good memory, all packed into a brain that in some birds is no bigger than a pea."[3]

How could we not be curious about how it works? Probably for as long as birds have flown off into the distance, those of us still rooted to terra firma have stood and watched, wondering where they might be going. And whether they might take a little piece of us with them.

People have been attaching things to birds for centuries. The ancient Romans sent messages with traveling birds like pigeons and swallows. Marco Polo, while traveling in Asia, noted small silver tablets attached to the feet of falcons. In Europe, herons released in Turkey with rings on their feet were later recovered in Germany. John James Audubon tied bits of silver thread on young phoebes' legs to better track and identify some of the birds on his Pennsylvania property. It wasn't until the late 1800s, though, that birds were banded for larger-scale scientific purposes. And as luck would have it, starlings played a starring role.

In 1899 a Danish schoolteacher named Hans Christian Mortensen cut up some strips of aluminum, bent them into rings, and affixed them to the legs of about 165 juvenile starlings. Each metal band bore an imprint with his address in Viborg and a unique number. He pinned his hopes on good citizens returning the bands and some information about where the birds had been found. Mortensen was so pleased with the results that he expanded his practice of ringing—as it became known in Europe—to ducks, storks, and birds of prey. The idea quickly caught on. German ornithologist Johannes Thienemann helped usher the practice into the scientific world, starting with his banding of starlings in Edinburgh, Scotland, in 1903.

Bird banding also took off in the United States. In one of the first experiments, Dr. Paul Bartsch of the Smithsonian banded twenty-three black-crowned night herons near Washington DC in the summer of 1902, each with the inscription "Return to the Smithsonian Institution." One report came back later that year of a heron that had been shot in Maryland, fifty-five miles away. He tried again the following year. Two years later he received a report about a heron found in Cuba with his band, apparently the first long-distance record of a bird banded in America. Percy Algernon Taverner, a self-taught naturalist and full-time architect, was so vexed by the questions of where birds went and how they lived that he furnished two hundred handmade aluminum bands, each with a message and unique serial number, to anyone ready to use them for science. That eventually yielded information on an intrepid flicker that traveled nearly nine hundred miles between Iowa and Louisiana. The American Ornithologists' Union organized the American Bird Banding Association in 1909. That year more than four thousand bands were distributed to forty-four different people, who affixed them to more than seventy bird species and then eagerly awaited word back. Bird clubs, museums, naturalist societies, rank amateurs, and ornithologists all got in on it. By 1925 the banders had their own published journal called *Bulletin of the Northeastern Bird-Banding Association*, later shortened to *Bird-Banding*. The banding frenzy was on.

The U.S. Biological Survey, which began as a small ornithological research operation at the U.S. Department of Agriculture, assumed much of the control of bird banding in 1920, keeping records of as many experiments as possible. By one estimate, more than four million birds were banded in America—and 331,000 reports were returned—between 1920 and 1944.[4]

It turned out starlings were good candidates for these tracking experiments. They were numerous, close to people, fairly easy to catch, and the subject of keen interest and derision. After all, isn't it always wise to know where your most annoying enemy is? "Bird problems are much easier to solve if information is available on the behavior of birds involved," federal biologist Olin Bray had said.[5]

EACH WINTER IN THE LATE 1920S THOUSANDS OF STARLINGS happily crammed into ten towers of Ohio's Columbus State Hospital, a sprawling psychiatric facility built in 1877 in the Kirkbride style, which emphasized decorative architecture, natural light, and grassy grounds. By then starlings were all over Ohio and developing a reputation as a nuisance. The uninvited birds, with their noise and their mess, ran counter to the serene atmosphere that the hospital administrators were trying to foster. But mysteriously, the starlings seemed to come and go with the seasons. So the question became, Where were the starlings when they weren't huddled in the hospital tower?

Ohio naturalist Edward S. Thomas and some colleagues had already been capturing starlings in barns in the central part of the state—apparently the first time flashlights had been used for that purpose—and banding them for release. More than twenty-five hundred had been banded before the crew came to the hospital. In just two years, more than four thousand squirming starlings were captured in nighttime raids at the hospital's elegant towers. "We ascend to the cupolas or towers by means of ladders, armed with bright flashlights," Thomas said. "The bright lights seem to befuddle the birds so that they flutter about in a bewildered manner and are easily captured. They are then lowered in gunny sacks to cooperators waiting below, who band and release the captives."[6]

With the work being relatively straightforward, save for the latenight hours and cold temperatures, the capturing-and-banding project only grew. Crews expanded from the hospital grounds to downtown Columbus, where birds were nabbed from warehouses, church steeples, and ventilation shafts. Outside of town, more than twenty roosts in rural areas were also looted. Over the course of seven winters, some thirty thousand starlings were caught and had steel bands affixed to their legs. Almost certainly, it was the most ambitious experiment of its kind to date.

Before the birds were released, though, most of them were sexed, weighed, and inspected for variations in beak coloration and plumage and any deformations. Lawrence Hicks, who worked as a biologist for state and federal agencies at the time, published an intensive set of detailed findings in 1934 about the variations among the starlings captured, noting that all parts of the state "had been invaded" by the birds, hinting it might be wise to become better acquainted with them.[7] But the question still nagged: Were these Ohio starlings migratory, and if so, where did they go?

It was well known then that in southern Europe, starlings tended to stay put year-round, content to weather whatever cool conditions came their way. In the north, though, starlings typically left in the winter, moving to warmer climes in the south. But it wasn't a straight north–south route. Unusually, they tended to move along a northeastern–southwestern line, diagonally across the continent.

In Ohio, migrating starlings exhibited a strikingly similar pattern, eschewing the typical north–south movement for a northeast–southwest path. The bird bands returned to Thomas and the other researchers revealed that while some starlings stayed within 20 miles of their winter redoubts, others traveled hundreds of miles. The farthest return came from Sainte-Marie, Quebec, some 700 miles to the northeast. To the southwest, one of their banded birds was discovered in Merigold, Mississippi, about 665 miles from its original capture. A map of all the recovered bands showed many birds were scattered between those two points on the northeast–southwest pathway, similar to the dispersal patterns of the starlings in northern Europe.

"It is a striking fact," Thomas said. Somehow the starlings in Ohio still carried ancestral information about the direction to fly during migrations, Thomas speculated, even though some of them had apparently not participated in migrations during their initial few generations in New York City. Something about the light or temperature in Ohio may have triggered those instincts to relocate. We now know that such information is passed down through generations of birds and other animals—developed over the countless generations looking for an edge in the competition for survival.⁸

But there was also this from the Ohio banding experiment: Some starlings migrated one year but not the next, upending the long-standing theory that migrating birds moved each year because of an innate and mysterious signal that required obedience. Why would they sometimes leave and sometimes stay? "This behavior is extremely puzzling," Thomas said.⁹

BEFORE SHE BECAME A PIONEERING ORNITHOLOGIST CHRONicling birds in nearly every corner of Alaska, Brina Kessel had spent her mid-twenties trying to untangle the riddle of migrating starlings in North America. She'd studied the birds for her PhD at Cornell University, publishing a paper about the best ways to tell a starling's sex and age, but couldn't quite walk away. In the spring of 1953 she offered the first comprehensive attempt to describe the movements of *Sturnus vulgaris* wherever it occurred in the United States.

The first thing Kessel noted was that starlings didn't behave in a uniform way. There were discernible differences even around Cornell's town of Ithaca, New York. Some marked starlings never left town, while others ventured as far as Quebec and Ontario. Even sibling birds banded from the same nest sometimes diverged in their destinations. In one case, one stayed home for the winter in Ithaca, while the other was found in western Pennsylvania. In another instance, one sibling flew off from the nest in Michigan to spend the winter in Wisconsin, while the other chose Missouri. "As a migratory species, the starling appears exceptionally plastic in its habits," Kessel said.¹⁰

It was a hard-won conclusion for Kessel, who had spent years sorting through thousands of records of starlings banded over the previous thirty years or so, including data on IBM computing cards kept by the U.S. Fish and Wildlife Service. To the untrained eye, the data seemed to show starlings flying willy-nilly, taking wing on whatever whim suited them best in the moment. Over time, though, patterns emerged from the data. Starlings, when they did move, tended to travel along valleys and plains, moving north in the spring and south in the fall. Within those rough parameters,

Kessel deciphered regional movement patterns, subtle changes in how starlings traveled.

In the Southwest and Texas, for instance, starlings came to winter but not to breed, and nearly all were migratory. Migrations were more common among starlings in the central and southern parts of the Midwest, while a larger number of starlings seemed to stay put around the Great Lakes, in the Northeast, and along the Atlantic Seaboard. Meanwhile, starlings on the East Coast seemed to move mainly north–south, while west of the Appalachian Mountains, they were likely to travel in a diagonal, northeast–southwest line. The youngest starlings tended to have the most variation in where and when they traveled, and their pioneering spirit took them to new territories as starlings moved across the continent, whereas the older ones generally followed a more fixed pattern.

Around the time that Kessel was sorting out the flying habits of starlings in the United States, one of the most famous starling migration experiments ever conducted was getting underway in Europe. Albert Perdeck, a Dutch biologist, was in his late twenties when he was put in charge of the Vogeltrekstation, a scientific institution based in the Netherlands that was dedicated to bird banding. By then it was fairly well known *where* starlings in Europe went during their migratory journeys. The question was shifting to *how*. In particular, were birds following some innate and mysterious sense of direction they'd been born with, or did they have to learn migration patterns from others? Perdeck, aiming for answers, decided to see what happened when starlings had their regular migrations drastically, but temporarily, interrupted.

Between 1948 and 1957 more than eleven thousand starlings on their fall migrations were caught in nets along the sand dunes in southern Holland. Left unmolested, these birds coming from the northeast would have continued south along the Dutch coast en route to warmer climes in southern Britain and northern France. But destiny, in the form of Perdeck and his crew, intervened. Those caught were quickly inspected, given a metal band, loaded into boxes, and then put on an airplane for Switzerland, some four hundred miles away. Once on the ground in Geneva, Zurich, or Basel, they were set free. Perdeck and his colleagues, curious about the fate of their birds, awaited the return of their bands. Ultimately, 354 bands came back.

When it came to analyzing the data from Perdeck's starlings, a split between adult and younger birds emerged. Once freed in Switzerland, the adult starlings were able to orient themselves, realize they were off course, and make a course correction to fly on a northwestern path to their typical destinations in northern France and southern Britain. The youngest starlings, though, which had probably never been on a fall migration, didn't make the adjustment. Instead, once let out of their boxes, they resumed flying south—much the same direction they'd been flying before they were caught on the Dutch coast—but now on a more southerly route than the adults. Many ended up in new places, like southern France and the Iberian Peninsula. The conclusion most often reached was that the adults were able to home in on their destination after having been there before, while young starlings were simply guided by a drive to fly on a fixed bearing, south in this case.

Perdeck ran similar experiments for decades, publishing a series of analyses from 1958 to 1983. The work was cited for years as evidence that his young, inexperienced starlings exhibited innate orientation when it came to migration, a sort of mysterious directional guiding system they'd been born with rather than learning the migration patterns from others in their flock.

But it hasn't become gospel in all ornithological corners. In 2020, for example, a group of Dutch and UK scientists published a complex critique of Perdeck's work, particularly examining how his results had been interpreted in subsequent years. They advocated a reexamination of what Perdeck's experiments actually showed and an acknowledgment that the results were likely more ambiguous than had been assumed. The scientists cited several factors that couldn't be ruled out as other explanations for Perdeck's results, including that the youngest starlings in the experiments may have been influenced by the topography when released or may have been following social cues from other starlings that hadn't been truly accounted for. Whatever the case, more scrutiny was needed on the simple conclusion that starlings were born hard-wired with some kind of magical compass. "The jury is out," they said.[11]

WHATEVER THE CASE, ALBERT PERDECK'S INVESTIGATION OF starlings on the Dutch coast was only partially instructive for those trying to understand the movement of starlings in the United States. Banding experiments continued at a brisk pace as regional researchers tried to understand where their starlings went during the year.

More than eight thousand were banded around State College, Pennsylvania, between 1950 and 1966. The returns confirmed earlier data noting the birds were moving along a diagonal northeast–southwest line by following topographical features like rivers and valleys.

Around the same time, researchers in Columbus, Ohio, decided to check the work of their colleagues from decades earlier. They trapped and banded more than sixteen thousand starlings during a single winter—and found the returned bands largely matched the migration patterns noticed by Edward Thomas and others in the 1920s and 1930s. There's something to be said for consistency.

In Colorado, more than twenty-six thousand starlings were caught, banded, and released during a fourteen-year experiment that started in 1960, with many of them also getting color-coded with plastic streamers attached to one leg. Most of the birds were captured at cattle feedlots and roosts in the central part of the state. While the majority of bands recovered came from within a hundred miles of where the birds had started, there were also some long-distance travelers in the bunch. Some showed up as far north as Manitoba and south into New Mexico and Texas.

Banding continues today, sometimes small batches and sometimes much larger. Nearly ten thousand starlings were scooped up and banded by the U.S. Department of Agriculture in the early 2000s in Texas, Kansas, and Nebraska. The scientists found that little had changed about the birds' movement patterns from the days when Kessel made her findings in the early 1950s. After nearly a century of banding starlings in the United States, there seemed to be scant new insight to be gained. "Starlings are renowned and persistent pests," the banding study said in 2012, reporting the obvious, "and the birds from our study areas probably contributed to agricultural and urban conflicts in several states."[12]

21

Poison Years

ON A WEDNESDAY NIGHT IN FEBRUARY 1975, NOT LONG AFTER dinnertime, the womp-womp-womp of military helicopters broke through the subfreezing chill hanging over the border of Kentucky and Tennessee. Suddenly, two Huey helicopters from the 101st Airborne Division came into view, lit partially by a nearly full moon, and pointed toward a stand of loblolly pines on the Fort Campbell army base where millions of starlings, blackbirds, and other birds were spending the night.

Veteran journalist Ann Cottrell Free was there and described the scene a few hours before the helicopters arrived, as the birds approached the pines:

> They covered the heavens there on the Kentucky-Tennessee border, not only overhead, but from horizon to horizon. We were in a world of blackbirds. Racing across the sky, they came at more than a mile a minute, swooping in flocks, turning and twisting as if one bird instead of thousands, soon, millions. Arriving from the foraging grounds up to 60 miles away, they came in battalions of dark beauty, guided, it would seem, by one great intelligence. Zooming into the grove, twittering excitedly, the flutter of wings brushing the air, they finally settled down for what should have been just another night before spring migration. But it was not to be.[1]

Over the next two hours the four-ton helicopters did battle with about five million European starlings, red-winged blackbirds, and grackles that had been parking themselves in the trees for years, much to the frustration of army commanders and neighbors of the base. The Hueys made twenty-two passes over the roosts, which occupied about eight acres of pines, and doused the birds with 160 gallons of water mixed with a chemical called Tergitol. It was a special detergent designed to dislodge the naturally occurring oil the birds had on their wings that helped keep them warm. Without the insulating oil, they would freeze to death in wet, frigid temperatures or starve, unable to eat enough to match the metabolic overdrive their bodies would need to try to stave off the cold.

After the helicopter spraying, with temperatures hovering around twenty degrees, two fire trucks were brought in to soak the birds with cold water, an attempt to wash off the natural oils for good and leave the birds as vulnerable as possible. They worked through the night, unleashing 120,000 gallons on the roosts before quitting just before dawn. "The chill weather did the rest, and the birds began dropping from the pine trees, catching on the branches and then tumbling to the ground, dead, not from freezing but from shock," another reporter, this one from the *New York Times*, wrote from the scene.[2]

Those that didn't die were left struggling on the ground, staggering as if they were drunk, clumsy and trying to stay on their feet. They couldn't fly, so they simply walked off, perhaps to die later or regain their body oils—and strength—in a few weeks.

Melvin Dyer, an avian ecologist from Colorado State University, was also at Fort Campbell that night, observing on behalf of the Society of Animal Rights, a New York–based group that had gone to court, unsuccessfully, to stop the spraying. He explained to the *Times* what seemed to have happened to the birds that died of sheer shock: "The nerve impulse stops. The heart stops. Breathing stops. Everything stops."[3]

The morning sun revealed the tally of the night's mission: about five hundred thousand dead birds scattered all over. More than a hundred soldiers from the army base were dispatched to collect most of the carcasses, stuffing them into plastic bags destined for the landfill. Others were left to rot in place.

While the night's work didn't wipe out as many birds as hoped, it was still one of the single largest wildlife-killing operations ever attempted in the United States at the time. Beyond the body count that night, the Huey raid opened a controversial new avenue for dealing with starlings, which had flummoxed wildlife control agencies for decades. The chemical war against this most hated bird, one that would ultimately wipe out more than eleven million starlings over the next nineteen years, had finally begun in earnest.

WHAT WAS SPRAYED AT FORT CAMPBELL WAS SOMETHING called a surface-active agent, or surfactant. Sometimes it's called a wetting agent. It's a substance that lowers the surface tension of water. When applied to the birds' feathers, it broke down their insulating oils and left them more exposed to the elements. It was like forcing someone outside on a frigid winter night without their clothes on.

The strategy of surfactants as bird control was first developed in 1958 at the Patuxent Wildlife Research Center in Maryland, which was then operated by the Bureau of Sport Fisheries and Wildlife. Biologist Dan Campbell noticed that both wild and captive blackbirds bathed in open water even during cold weather. He theorized that ridding them of their insulating oils might leave them vulnerable to freezing. Several tests in which caged birds were hand-sprayed with detergents were run at the federal center in Maryland and another in Florida. The results were promising (at least for the researchers), kicking off a frenzy of experiments.[4]

Tests were conducted in the ensuing years around the country, first in labs and then in the field. In 1966 a roost of birds in Kentucky was rousted and forced to fly through a curtain of chemicals sprayed from nozzles mounted to a ledge. About twenty thousand birds were killed in just over twenty minutes. Similar experiments were tried in Ohio, Michigan, and New Mexico. Elsewhere, three trials at a holly orchard, where floodlights attracted birds to fly through a set of vertical strings doused in the detergent, killed eighty thousand to ninety thousand starlings.

Ground-mounted sprayers were one thing, but what about simply dumping the detergent on them from the air? Maybe something military grade?

Teaming up with the U.S. Air Force, researchers borrowed a C-123 airplane, the same model often used to spray Agent Orange during the Vietnam War, to dump 950 gallons of mixture on a blackbird roost at Georgia's Moody Air Force Base. When that didn't yield great results, they switched to a B-26, a twin-engine cigar-shaped bomber used frequently during World War II. Again the roost at Moody was targeted, only this time researchers also placed a caged sunfish in the drop zone to test whether the surfactant would be harmful to fish. The sunfish survived the dousing and seemed to fare about the same as a caged sunfish set up elsewhere as a control, but several thousand birds died after two drops of a thousand gallons of the chemical mixture. The bomber was then sent to Arkansas to run similar experiments on a giant roosts of starlings and blackbirds. Over the course of seven detergent drops, some seventy-eight thousand birds were killed, including about twenty thousand after a single drop. Some of the deaths happened when they were doused; other birds died afterward in the rain or when they were sprayed with water. The researchers soon upgraded to using a PB4Y-2, a large patrol bomber used during World War II and the Korean War. It was capable of dropping two thousand gallons of solution.

The idea was starting to take shape, but researchers were still unclear about the best detergent-to-water ratio. Was it better to dump a lot of water with a small fraction of oil-scrubbing soap or less water with a higher concentration of the chemicals? Hundreds of thousands of starlings and blackbirds were killed during the next round of experiments in Alabama, Ohio, and Arkansas. The death count went up when it rained after the spraying and when the temperatures remained cold. Some guidelines emerged: douse the roosts during winter nights, preferably when rain is on the way.

The spray, officially known as Compound PA-14 Avian Stressing Agent and sometimes called Tergitol, was approved by the U.S. Environmental Protection Agency in 1973. At the time, PA-14 was the only lethal control the agency had approved for starlings and blackbirds. For people feeling under siege from the birds, it couldn't come on the market soon enough. By the winter of 1974–75 there were some 537 million roosting blackbirds and European starlings in the United States. And almost always, the starlings seemed to produce the most exasperation.

"Indeed, they seem as hard to kill as Rasputin," Rep. Frank A. Stubblefield of Kentucky told his colleagues in December 1974 during a congressional debate over starlings. "We must rid ourselves of this health and economic menace. Eradication seems to be the only answer." And what better place to try out the new weapon than the tenacious flocks that had been amassing along the border of western Kentucky and Tennessee?[5]

STARLINGS, BLACKBIRDS, GRACKLES, AND COWBIRDS HAD BEEN in the area for years and didn't cause much of a stir. By the late 1960s, though, the growing gatherings in the pines at Fort Campbell—which began every year with the birds' arrival in the fall and ended in late winter or early spring—started to become a problem. By the early 1970s it had blossomed into something of a crisis.

During the day, the birds descended on the fields of surrounding farms, feasting on wheat and barley and causing havoc among the animals. When the sun went down, the birds retreated to Fort Campbell and the safety of the trees, a pattern they repeated throughout their stay. The birds were sometimes so numerous in the air that military flights had be delayed when the birds were traveling at dawn or dusk, not ideal for the army's premier air assault base, home of the Screaming Eagles of the 101st. The vast traveling clouds of birds were both a danger and a spectacle. "Awesome is the word for it," said Col. Robert Peach of the 101st.[6]

But it was the effects beyond the military borders that really chafed. By 1975 there were an estimated five million starlings, blackbirds, and others at the Fort Campbell roost, and some seventy-seven million living in the hundred-square-mile area around it. That included about twelve million birds in Christian County, Kentucky; twenty million in Logan County, Kentucky; another twenty million in Milan, Tennessee; and some eight million in Montgomery County, Tennessee.

In the winter of 1973–74 it was estimated the starlings and other birds had done $5.8 million in damage to local crops and livestock. Not only were there widespread complaints about sick pigs and devoured grains, but the birds were also potentially linked to people becoming sick from a lung ailment called histoplasmosis. The disease came from a fungus that lived in the soil, particularly soil with a lot of bird or bat droppings. When stirred up, the soil released spores that could then be inhaled by people and cause flulike symptoms that could send victims to the hospital. The fungus was common in the soil of that part of the country, so it could be hard to distinguish which human cases were directly related to the bird roosts in the area and which weren't. Still, local officials said there had been a significant increase in the disease since the birds' arrival. In Christian County, Kentucky, there were two cases of histoplasmosis in 1970 and twenty-one in 1974, according to a local health official.

"These birds have literally taken over the communities in which they are present," Kentucky senator Wendell H. Ford said in the winter of 1976. "The effect has been devastating on the lives and livelihoods of the people who reside there. Farmers have seen their feed and grain crops destroyed. They have borne the loss of livestock due to the disease carried by these birds. And they have watched helplessly as this disease—histoplasmosis—spreads to their families and friends." As governor a few years earlier, Ford had declared a state of emergency over the birds. The governor of Tennessee had done the same. People were scared, frustrated, and tired of the fruitless attempts to drive the birds away. "We have tried every resource of which we know; and we have failed," Ford said.[7]

Residents of Hopkinsville, Kentucky, formed the Bird Reduction Committee, which was soon flooded with suggestions on the best ways to deal with the unwanted birds. "We get suggestions all the time to set up a blackbird canning factory," said Tracy Carter, the town's ombudsman. Other ideas included using rubber snakes, ceramic cats, or fake owls; placing electrified wires in the branches of the trees; and parking cars beneath the roofs, gunning the engines, and training the fumes onto the birds. The town eventually paid a bird control expert from California $10,000 to play recorded distress calls from loudspeakers mounted on a helicopter. The birds didn't budge. "He didn't scare a single one," the mayor said.[8]

The fight against the birds started to make national headlines, and the conversation soon turned to using Tergitol, or PA-14, to kill them or at least make a dent in their numbers. Some conservation groups began raising concerns about the use of chemicals for widespread killing. Because those groups were based on the East Coast, the locals in Kentucky and Tennessee railed against them. "If we tried to save their rats the way they try to save our birds, they'd blow the roof off. You'd never hear the last of it," said James Simmons, a farmer in Russellville, Kentucky who said the birds had descended on his seven-hundred-acre farm, devouring his grain seeds and making his young pigs get sick and die.[9]

It was a common refrain: if the city slickers in New York were able to kill rats in the name of pest management, rural folks ought to be able to do the same with the starlings and the blackbirds in western Kentucky and Tennessee. And not lost was the fact that the starlings had been introduced into Central Park by aristocrats nearly a century earlier.

"If we could ship 14 million starlings to the Pentagon, or Central Park, we'd get results," said George L. Atkins Jr., mayor of Hopkinsville. "It's a pestilence and a scourge. Farmers are in the fields with shotguns, cattle and hogs are driven from the feed lots, children's slides are covered with bird droppings." At one point a furious Atkins had even asked city attorneys to draft a request for a legal injunction to stop New York City from killings its rats. "That's about as inhumane!" he railed. "That's the absurdity of it!"[10]

AFTER THE 1975 SPRAYING, IT BECAME CLEAR THAT MANY OF the surviving starlings, blackbirds, and grackles were staying put at Fort Campbell and that the remaining birds in the region had no intention of leaving either. Soon politicians and locals were pleading with Congress to step in and approve a bill providing emergency authorization to deploy PA-14 in the two states.

"Several years ago, the master of movie thrillers, the great Alfred Hitchcock, fictionalized a situation in which millions of birds terrorized humans, preventing them from leaving their homes, and in the end, mounting deadly assaults on the human population," said Tennessee representative Robin Beard. "While no one would presume to say that these present-day birds are plotting an organized attack, in all other respects, the same situation fictionalized by Mr. Hitchcock years ago exists today in portions of Tennessee and Kentucky."[11]

On February 2, 1976, a daylong congressional hearing was held on the emergency bill in hopes of passing it that day and sending it along for President Gerald Ford to sign immediately so that the birds could be sprayed with the detergent before winter ended and moved along for the rest of the year. Not everyone who testified that day shared such a nightmarish vision of what was happening in Kentucky and Tennessee.

Two groups, the Society for Animal Rights and Citizens for Animals, had sued to halt the spraying of Tergitol in early 1975. They managed to get it delayed by eleven days before the U.S. Supreme Court denied a plea to take up the case. The groups were back in court in September 1975 to get an injunction against more spraying at Fort Campbell, saying the government hadn't done enough to investigate the potential long-term impacts as well as alternatives that might be suitable. In negotiations that fall, the government agreed to do a more comprehensive environmental review, but an agreement was also struck allowing Tergitol to be used if there was a substantial and imminent harm to people from a specific roost and if all other nonlethal means had failed.

Bruce Terris, an attorney for the Society for Animal Rights, argued that the evidence provided by the two states had utterly failed to definitively tie the cases of histoplasmosis to specific birds roosts, noting that the fungus could also be stirred up in chicken coops and other places. He maintained that locals were scapegoating the birds for a disease that had plenty of other local sources, all in pursuit of permission to try to permanently eradicate birds they saw as a nuisance. "I think these public officials that represent the people in Kentucky and Tennessee do them no service by misleading them into the . . . use of Tergitol to kill indiscriminately millions of birds and dozens of roosts," Terris said. He asserted that state officials seemed determined not to pursue other, less lethal options, like bulldozing the sites of large roosts. All the while, they ignored any possibility that starlings or blackbirds might actually do some good. "They are among the birds that consume the most insects of any in our bird populations and so it is not obvious by any means that these birds are, on net balance, harmful to the people of this country."[12]

Ann Cottrell Free, the writer who had been at Fort Campbell during the spraying in February 1975, said she could find no evidence of histoplasmosis cases that could be linked to the birds. She, too, accused elected officials of masterminding what she called "Operation Hysteria," a ploy to annihilate millions of birds and perhaps boost the careers of a few politicians along the way. "The political leaders, looking ahead to the next elections, no doubt, sought to please their constituents by slaying the dragon," Free said in an article that was submitted as part of her testimony at the hearing, "even if they had to pump a little fire and venom into it first."[13]

But not everyone in the environmental community stood against the use of Tergitol in the area. The Audubon Society in particular found itself caught in the middle, with bird lovers divided over what should be done. "Those against the killing say, 'Even starlings have a right to live. It's not their fault they were imported from Europe,'" one local Audubon newsletter said by way of explanation. "While those for the killing say, 'In such numbers the birds are a health menace and anyway starlings are not native and they are crowding our many native, more beautiful birds.'"[14]

While remaining conspicuously silent in public, the group's national leaders sent a statement to their chapter organizations to explain their position. The gist was that while the organization "deplored the recent slaughter" at Fort Campbell, it reluctantly didn't oppose it. Fort Campbell had wrongly allowed the bird numbers to flourish and expand by allowing fifteen thousand acres of roost-friendly pines to grow in a place where hardwoods had typically dominated. The birds posed a danger and something had to be done, Audubon officials said. "Don't misunderstand our position. Fort Campbell was an exception and an emergency. We continue to believe there are better ways to handle wildlife problems when—and if—such problems actually exist."[15]

Floyd Ford, a biology professor at Austin Peay State University and a member of the local chapter of the Audubon Society, said at the congressional hearing in February that while his local group opposed poisoning the birds, they did support use of the detergent. The influx of millions of birds, especially starlings, had driven down the number of local songbirds and left the ecosystem out of balance. One shouldn't worry too much about the fate of starlings, he said, noting that these "pests" were perhaps the hardiest survivors in the bird family in North America and could withstand the losses that might result from the bill expanding Tergitol's use. He invited "any persons . . . who have not seen with their own eyes the extremely serious situation" to come to his home in Clarksville, Tennessee. "I will personally drive them to as many of the roosts as they care to see," he said. "My guess is that only one, or two at the most, will be sufficient."[16]

THERE WAS NO NEED FOR MEMBERS OF CONGRESS TO TRAVEL far to see starlings in action. Just a few blocks down Pennsylvania Avenue, huge quivering flocks perched in the magnolias on the South Lawn of the White House. A few years earlier, President Nixon had complained about the noise of the starling infestation outside his windows. President Ford inherited the annoying birds when he took office.

A man named Dale Haney was given the full-time job of trying to shoo them away. "During the day, they feed in the countryside around Washington," Haney told the *New York Times* the same month as the hearing over the PA-14 bill. "But they come swarming in here at sunset. It's horrible." Haney had tried all the tricks, like coating the trees with a sticky goo, blasting recordings of bird distress calls, and banging like mad on pie pans beneath the branches of the magnolias. Had anything worked? "No," a defeated Haney said, "there's no way to beat them."[17]

To no one's surprise, Ford signed the bill to start emergency spraying of PA-14, first in Kentucky and Tennessee, and soon far beyond. Over the next two decades more than eighty massive roosts were doused with the detergent. Government officials later estimated the operations wiped out more than 38 million birds, including 11.4 million starlings. The government stopped using PA-14 in 1992, deeming it too costly to provide the data that had been requested by the Environmental Protection Agency about its long-term effects. But America's forever war on starlings had entered a more technical, and far more lethal, new phase.[18]

22

A Forever War

ON THE WESTERN EDGE OF COLORADO STATE UNIVERSITY, UP a nondescript road on the outskirts of Fort Collins, the headquarters of the U.S. Department of Agriculture (USDA) National Wildlife Research Center sits in the foothills of the Rocky Mountains. The forty-three-acre spread is hard to see from the road, and maybe that's no accident. The property is surrounded by a high-security fence, the kind you might expect at a military installation. When I arrived early on a summer morning, a security guard met me as I approached the gate. He took my ID, had me sign an oversize ledger, asked me about my appointment, and then cocked his eyebrow in a way that hinted perhaps the place didn't get many lone visitors.

The research center is the latest iteration of the lab started by Edwin Kalmbach in Denver back in the 1940s. Now, instead of being part of the U.S. Fish and Wildlife Service, it's operated under the USDA's Animal and Plant Health Inspection Service. It's the research arm of the USDA's Wildlife Services program, which was established decades ago to reduce conflict between wildlife and people, including at airports, farms, ranches, cities, and suburbs. Although the USDA still tries to control native species like coyotes, beavers, and mountain lions, much of its emphasis now is on nonnative and invasive species. The United States has been inundated with foreign species over the past 150 years or so: kudzu, zebra mussels, Russian thistle, feral hogs, brown snakes, Burmese pythons, carp, cane toads, lionfish, giant African land snails, Japanese beetles, giant hogweed, scotch broom, sea lamprey, starlings—the list goes on and on. To varying degrees, each threatens native wildlife, the environment, people, or property, and each poses a vexing problem in search of a solution. It's an uphill battle made more difficult by a number of factors: people and animals moving around the globe like never before, Darwinian forces that often favor expansive and aggressive foreigners over native species, the rise of industrial agriculture, emerging diseases and vectors, wildlife trade, habitat destruction, rampant consumerism. Again, the list goes on and evolves by month and year. It's a mind-boggling task to keep up with the latest developments, much less try to engineer solutions.

About 150 people work at the research center or one of its eight field stations scattered across the country. Years before I visited Fort Collins, I stopped by a research facility in northern Utah that was focused almost exclusively on figuring out how coyotes and people might coexist a little better. Coyotes are every bit as adaptable and tenacious as starlings, though they're native to North America and I'd argue warrant more tolerance, since they were here first. The Fort Collins facility doesn't have the luxury of focusing on just one or two of the toughest problems. Baffling questions abound: How can we keep five million feral hogs in thirty-five states from tearing up farms, ranches, and suburban neighborhoods? How can we stop introduced brown tree snakes in Guam from driving native birds and bats into extinction? How do we best protect sheep and cows from bears, wolves, and coyotes? How can we stem the spread of diseases like West Nile virus, SARS, rabies, and chronic wasting disease? How can we protect native species and people from pests gone wild like nutria and pythons?

Because the degree of difficulty of these questions is so high, all manner of science is deployed at the research center. During a walk-through with one of the staffers, I heard about genetic research, epidemiology, molecular and cellular biology, reproductive physiology, virology, chemistry, economics, ecology, chemistry, surveillance techniques, and statistics. And of course, the facility also relies on live animals, which are sometimes kept in outdoor pens, inside cages, or special interior rooms that can be outfitted to simulate natural environments like temperate forests or tropical swamps. The rooms were clean and eerily quiet when I visited, and the hallway air seemed scrubbed with the kind of benign smell designed not to be noticed. At times, though, the research campus can feel like a menagerie holding a who's who of troublesome (to some) animals. A hall sign listed a sampling of species that had been housed at the center, including crows, woodpeckers, pelicans, geese, prairie dogs, bats, rabbits, raccoons, pigs, trout, bullfrogs, and snakes. At one point, ground squirrels were brought in as researchers wrestled with developing special fencing to keep the tiny, two-pound rodents from tunneling into a nuclear missile base in western Montana.

Starlings are at the research center a lot, and for good reason. The USDA estimates they cause $1 billion in damage to agricultural operations every year, making it probably the most costly bird species in the United States. In 2012 the agency figured starlings cost farmers $189 million just by digging into five fruits: blueberries, wine grapes, sweet cherries, tart cherries, and apples. That doesn't include other crops, like sweet corn and grains. At a large livestock operation, a flock of a thousand or so starlings can eat 1.5 tons of cattle feed in two months—a scenario where each individual starling might be costing the operation $200 to $400 during that time. In a city, the mess the starlings leave behind can be expensive to clean up. The USDA estimates that a roost of about thirty-five thousand starlings in a downtown setting might cost a building owner $260,000 in cleaning and maintenance over two years. More difficult is putting a dollar figure on the cost of their spreading diseases, hindering native birds, or necessitating the starling-specific work that goes on around airports.

Scott Werner, a research wildlife biologist, has spent more than twenty years working on starlings, often trying to find ways to keep them out of corn and rice crops, away from dairy farms, and out of the way of airplanes. Over the years, he's had hundreds of starlings in his labs, which tested repellents, chemicals, and even pellets that are milled at the research center for experimenting and use in the field.

One recent project focused on ways to reduce the amount of starling droppings that fall on farm equipment, buildings, and walkways. The feces can be acidic and corrosive, as well as a slip-

and-fall hazard for workers. Werner and his team evaluated several kinds of repellents—each with its own chemical compounds—that could be applied to perches above those target areas. Some were sticky and others slippery. All of them aimed at being unpleasant enough for starlings to force the birds to look for somewhere else to rest, while being safe enough for the environment. Another research project, this one with Cornell University, looked at patterns of genetic diversity among starlings captured at dairies and feedlots in seventeen states. The hope was to identify factors that had allowed starlings to adapt and spread so quickly. There were few genetic differences, but some of the gene sequences correlated with certain temperatures and rainfall—a potential clue for understanding how local adaptations evolved so fast. That work would have to continue. Earlier, Werner and others ran a series of experiments looking at whether expanding the size of food pellets given to dairy cows might stop starlings from raiding the feed bins. They found that making the pellets slightly larger reduced starling consumption by 79 percent. He's also analyzed roosting behavior, interactions with cows, new formulations of chemical repellents aimed at keeping starlings off blueberries and sweet corn, and better ways for frightening birds.

Starlings are complex, and the problems they pose are multifaceted and situational, Werner said. They also possess a plasticity and dynamism that makes it difficult to stay one step ahead of them. Drop them almost anywhere, and chances are they will find a way to survive. "These animals make a living staying alive and they're very good at it," Werner told me, offering some grudging admiration. "They're doing a better job than human beings of exploiting their environment."

Alan Franklin, another longtime researcher at the Fort Collins facility, offered his own blunt take on starlings. "When the human race is dead and gone," he said, "starlings will be around along with the cockroaches and Canada geese."

Much of his work examines starlings as a vector for diseases like salmonella, avian influenza, and listeria, as well as E. coli. A lot of birds can carry diseases. What sets starlings apart, and what can be so troublesome, is that they travel in such large numbers, are highly mobile, very adaptable, and capable of being tenacious survivors in so many kinds of environments. "That's a recipe for disaster when you're talking about disease transmission," Franklin said.

There's been a particular focus in trying to understand how starlings might convey disease in concentrated animal feeding operations (CAFOs), outdoor industrial-scale livestock operations that often include thousands of cows in tight spaces, especially when clustered together for food and water. CAFOs have proliferated over the past several decades, and starlings have arrived right along with them, often ready to feast on the food put out for the cows. Franklin and other scientists have worried about how disease-causing bacteria like salmonella and E. coli might spread rapidly in CAFOs, move through the food chain, and find their way to people. Birds drawn to CAFOs are certainly part of the mix, and in Franklin's mind, starlings are the most troubling culprits for their sheer numbers and ability to easily transport bacteria, viruses, and diseases.

Yes, researchers should be looking at ways to repel starlings and keep them away from livestock operations, Franklin said, but it's not enough. The United States should be looking at a national strategy for eradicating starlings—an approach that will drive them out and eliminate them forever. It's a tall order at this point, Franklin admitted, but that doesn't mean it doesn't deserve consideration, given the difficulties over the past century. "I don't understand why we're not doing that," he said. It's not as if some aren't trying.

IN JANUARY 2009 THOUSANDS OF DEAD STARLINGS FELL FROM the sky, landing on frozen roads and lawns in the rural community of Griggstown, New Jersey, not far from Princeton. Residents were alarmed. "It was raining birds," the mayor said. "It got people a little anxious." A couple of years later more than two hundred dead starlings turned up in snowy downtown Yankton, South Dakota. It happens periodically around the country: a scattering of starling corpses show up, seemingly overnight, and leave residents a bit weirded out and wondering what's just happened.[1]

The mystery is typically solved pretty quickly: agents with the USDA's Wildlife Services program have taken measures to rid a nearby farm or agricultural operation of a persistent flock of starlings. The agency, unique in the federal government, is tasked with doing daily battle with wildlife across the country. Its roots go back more than a century to a time when the government was asked to help eradicate bears, mountain lions, wolves, coyotes, jaguars, and other species classified as a threat to people or livestock operations. Today the work has vastly expanded but often happens out of the public eye—a campaign of harassment, redirection, or killing designed to protect people, property, and profits from wild animals that are deemed troublemakers.

The reasons are as varied as the species. Some are targeted because they like to snack on calves or sheep. Others are a nuisance around airports or have wandered into suburbia or dug burrows in places where people think they shouldn't. Some might spread disease or foul water supplies. Still more are nonnative species threatening to disrupt fragile ecosystems, drive native species toward extinction, or feast on crops. Wildlife Services stays busy. In 2020, for example, the agency reported killing more than 62,000 coyotes. That year agents also killed 434 black bears, 276 mountain lions, 685 bobcats, 2,527 foxes, and more than 25,000 beavers. The work isn't without controversy, not just for the large number of animals killed but also for the way it's carried out. Wildlife Services kills in a lot of different ways, employing poisons, bone-crushing traps, explosives, and sharpshooters in airplanes. Occasionally, the program's operations generate newspaper headlines, and there's a flurry of public discussion about whether reform is needed. More often than not, the controversy passes and the program soldiers on.

But no animal is targeted more often, and in such abundance, as the European starling. No other species really comes close. I combed the agency's records going back decades and found that between 2008 and 2022 Wildlife Services killed more than 18.5 million starlings. That's about 1.2 million per year, on average, or more than 3,300 every day of the year. During that same period, the agency said it chased off more than 161 million starlings (often more than once, it's safe to bet) from airports, city roosts, agricultural operations, suburban gatherings, and anywhere else they showed up. On average, that means 29,000 starlings were dispersed every day.

Much of the killing happens at CAFOS, the superbig cattle operations that starlings can't seem to quit, especially when other food sources are scarce in the late fall and winter. The agency's preferred starling poison remains DRC-1339, the slow-acting chemical developed between federal researchers and Ralston-Purina back in the 1960s. It comes in two forms: a pale-yellow crystalline powder that's soluble in water and poisonous bait pellets once registered under the trade name Starlicide. It can be easily mixed with oats, raisins, or french fries and left out for starlings. A technique called prebaiting is often used, in which nonpoisoned food is left out to acclimate and entice the starlings before the poisoned versions are swapped in. Once they ingest DRC-1339, most die within three days, typically of blood poisoning due to kidney malfunction. But it's not so easy to control where they finally succumb, hence the starlings falling from the sky in places like rural New Jersey and downtown Yankton.

USDA officials like this poison because it can be used to specifically target a particular species, with doses carefully measured out to deal with the species and situation at hand, and it poses relatively few problems for nontarget species, including other animals that might eat a poisoned starling. One session with the poison can wipe out 70 to 100 percent of a flock if there are only ten thousand to twenty thousand birds. Any more than that and more work will need to be done. It's used on other birds, too, including pigeons, grackles, crows, ravens, magpies, and gulls, but starlings are the most frequent recipients. DRC-1339 is Wildlife Services' most common way to kill—more than 70 percent of all animal deaths in its work can be attributed to this product, according to a review published in 2022. As effective a killer as it is, it's still not a cure-all.[2]

The USDA's Wildlife Services today offers a menu of suggestions for dealing with starlings, none of them perfect, all of them providing a modicum of temporary relief, much like Edwin Kalmbach's suggestions back in the 1940s. Starlings still become habituated to distress calls or objects like kites and fake owls. Netting can be effective at keeping them off some crops, but it can be expensive. It's the same with specially built covers. Traps that lure in starlings can sometimes work, especially with unsuspecting juveniles, and then the birds need to be killed. A box full of carbon monoxide can do the trick. Cutting down or thinning trees can help reduce roosts.

Pyrotechnics, scarecrows, and battery-operated alarms can sometimes work. Perhaps better are propane exploders, which can be mounted on the ground or in the back of a truck. Set one off every five or ten acres where the starlings are, and they're apt to at least move for a time. Bait infused with a chemical called Avitrol can mess with starlings' minds, making them anxious. Some will act erratically, even emitting shrieking calls that freak out the other starlings. Sometimes the drug might kill them. There are sticky repellents that keep starlings from roosts, but they often wear out. Other repellents are designed with tastes to turn off starlings or upset their stomachs, with varying degrees of success. And then there's good old-fashioned shooting. But, the agency warns, it's "generally not an effective damage management technique."[3]

There's a good reason starlings have hung around this long, and anyone thinking they might simply eradicate starlings faces stiff headwinds from math and biology. Ornithologists say most starlings live three to four years, and many breeding pairs produce two clutches of eggs each year, with an average of five eggs each time. Even though most young birds don't make it to adulthood, the math still works in their favor, with more being born and fledging than the sixty million to seventy-five million that die each year from natural causes like starvation, predation, and disease. Even the federal government's starling-killing programs, which might top three million some years, doesn't do much to curb the population.

Years before I came to the Fort Collins facility, I talked with a retired Wildlife Services researcher named Richard Dolbeer, one of the top experts in the country when it comes to thinking about ways to deal with starlings. He lived in Sandusky, Ohio, and was well versed in starlings' presence there and everywhere. We ran through the staggering array of options that the USDA and others have at their disposal. The best approach, he said, was an integrated one that combined protecting spaces, shooing birds, and killing them when necessary. But expectations ought to be managed. "It's sort of like bailing the ocean with a thimble," he told me.[4]

BEFORE I LEFT THE NATIONAL WILDLIFE RESEARCH CENTER in Fort Collins, I spent several hours in the archives, digging through box after box of documents about starlings. I found field reports, private memos with scribbled notes, internal arguments over whether starlings should be eaten, criticisms of the Bird Men like Otto Standke, and even several of the record albums from the 1960s containing starling distress calls. Had any bird generated so much research, so many harebrained schemes, such distress and consternation? Woven through so many of the documents was a thread of deep frustration and futility. It left me wondering: How had a three-ounce bird, purposely invited onto our shores, gotten the better of us?

On a shelf not far from the archival boxes, I spotted a blue and orange cannon, one of those noisemaking propane exploders. This one, which looked to be decades old, was made in Belgium and labeled Thunderbird Scare-Away. It was worn and well used by the time it had been retired. How many explosions had rattled through its muzzle, carrying hopes that it might finally shoo away the starlings from someone's farm? How soon was it before the frightened birds, flushed into the air by the booming noise, returned to their roost, barely perturbed at all? This propane gun was shelved now, silent and idle. Outside, the starlings carried on.

23

Spellbound

OFTEN ON COOL EVENINGS BETWEEN NOVEMBER AND MARCH, thousands and sometimes millions of starlings come to Ham Wall in southern England, a nature preserve overseen by the Royal Society for the Protection of Birds. The starlings, passing through on their way south, roost in the Avalon Marshes each night. But before settling down at sunset, they flock together and put on a spectacular aerial display, swishing across the sky in mesmerizing designs. People, too, flock to the area to catch the show. Occasionally, more than a thousand starling watchers show up, some describing it as a deep spiritual experience. It's become so popular that the preserve had to add extra parking, toilets, and a snack bar. They even set up a "starling hotline" to call ahead of time to get the best sense about where the starlings are likely to appear.

Even the staunchest of starling haters has to admit that when they're in the air, swarming in great gyrating shapes formed by thousands of individual birds, they're a bewitching sight. You can sense a pattern in the murmuration, but it seems determined to remain elusive. "Part of the fascination of the starlings is the way they seem to be inscribing some sort of language in the air, if only we could read it," author Jonathan Rosen wrote.[1]

Many look like little more than black smudges in the sky at first, but then forms and shapes emerge and shift and suddenly vanish to become something else entirely, transforming and retransforming, never ceasing to transfix and mystify. One two-minute video, shot by two young women canoeing in Ireland and featuring a beguiling surprise show of starlings over a lake at dusk, generated more than ten million views within a few months of being posted. "I cry everytime I watch this," one viewer commented.[2]

For a time, around 2015, swirling clouds of some thirty thousand starlings appeared at dusk above a suburb north of London. Locals set up chairs in their gardens to watch, children on bikes stopped on street corners, and traffic slowed to a crawl while the show unfolded overhead. Everything seemed to come to halt for the nightly ritual, everyone marveling at the spontaneous expression of nature in the air above their humble neighborhood. People swapped tips about the best places to view the starlings and speculated about what they might be doing up there. "You can hear the noise of all the wings beating," one witness said. "At times they appeared like a swarm of bees, other times like a truly massive swirling cloud. They turn and swoop, sometimes splitting into two or three groups. Then the separate groups come back together like colliding galaxies."[3]

Danish photographer Søren Solkær spent three years following migrating starlings across Europe, capturing them on film and video in England, Italy, Ireland, Catalonia, and the northern reaches of the Wadden Sea along the shores of the Netherlands, Germany, and Denmark. "Here, each spring and autumn, the skies come to life with the swirling displays of hundreds of thousands of starlings—an event known locally as 'sort sol,' or 'black sun'—as the birds pass through on their seasonal migrations," he wrote in a piece for the *New York Times* in 2022.[4]

He compared their swirling murmurations to a brushstroke in calligraphy, sometimes meditative and other times dramatic "as they perform a breathtaking ballet, one with life-and-death consequences," he wrote. "At times the flock seems to possess the cohesive power of superfluids, changing shape in an endless flux. From geographic to organic, from solid to fluid, from material to ethereal, from reality to a dream: This is the moment I attempt to capture—a mere fragment of eternity."[5]

It's no wonder ancient Romans sent priests out to stand beneath these giant, shifting swarms and divine significance. Or that

Edmund Selous, the British naturalist who spent so much time watching starlings in England's Suffolk County in the early 1900s, concluded that the birds operated with some kind of starling-to-starling telepathy as they whipped around the sky, dissolving from one fantastic formation into another again and again. The truth is that we still don't know exactly why starlings cluster together in the sky before parking in a roost before sleep. Maybe they're trying to expend some of the day's last energy before night falls. Or maybe the flamboyant flocks help gather any stray members that may have wandered off during the day. More than likely it has something to do with avoiding owls, hawks, and other hungry predators. Whatever the motivation, the execution remains a thing that suggests discipline and hidden organization. "Flocks are often observed to execute aerial evolutions with the unanimity and precision of a company of well-drilled soldiers," ornithologist May Thacher Cooke observed a few decades after starlings arrived in America.[6]

More recently, ornithologist Noah Strycker described it like this in *The Thing with Feathers*: "Starling flocks, it turns out, are thinner than you might expect—more like a floppy pancake than a football. The pancake slides around in various directions, shifting its appearance, but generally stays parallel to the ground and maintains a constant proportional shape, no matter the size of the flock."[7]

Typically, he said, there are more birds packed on the outer edges of the flock and fewer in the middle. "And starling flocks don't have leaders. When a flock turns, birds fly on equal-radius paths; in other words, they each turn on the same curve at the same speed." He compared it to a column of marching soldiers where those on the outside must walk faster while on a turn in order to keep a straight line. But, he added, "starlings don't compensate like soldiers do, so birds at the front of the flock end up on the right side after a left turn, those on the right side end up at the back, and those in the back end up on the left side."[8]

So then, as the flock shifts and moves, no bird is constantly stuck in the front position or on the outer fringes, where they're more vulnerable to being picked off by a predator like a hawk. That may play an important role in fostering cohesion within the flock, Strycker said. "Any bird forced to spend all of its time on the edge would be less motivated to stay in the group, and the whole arrangement could fall apart."[9]

Starlings' murmurations aren't all that different from schools of fish or swarms of insects that seem to be governed by some unseen force—a kind of collective brainpower, a sum more powerful than its individual parts—that tells them when to zig and when to zag. But how do the starlings know when to turn, wheel, and bank in the sky? And how do they keep from bumping into each other?

IN 1956 *PSYCHOLOGICAL REVIEW* PUBLISHED A PAPER BY George A. Miller, a cognitive psychologist at Harvard, titled "The Magical Number Seven, Plus or Minus Two: Some Limits on Our Capacity for Processing Information." The most common interpretation of the much-discussed paper is that the average person can hold only seven objects in their short-term memory—words, letters, numbers, items—plus or minus two. As it happens, starlings may share a similar limitation that informs their collective movements in the sky.

In 2007 physicists and statisticians in Rome set up cameras on a terrace at the Palazzo Massimo, a palace that was converted decades ago into the National Roman Museum. The cameras were pointed toward the sky in the direction of a railway station where a huge flock of starlings, after spending their days in the countryside, liked to roost each night. Before settling down each evening, though, the birds swirled in great flocks above the city at dusk, entrancing visitors and residents alike while also irritating many with the mess they left behind. For years, the researchers went up on the roof of the museum to capture data from the nightly aerial displays. Their high-resolution cameras, set about twenty-five meters apart from one another at different angles, were linked and took ten simultaneous frames per second when the birds showed up. The researchers were trying to do something that had never been done: create a three-dimensional model of starling flocks in hopes of learning more about how they moved and why.

Tracking collective behavior requires an understanding of each individual's behavior. But in Rome, once the photos were in a computer program, the difficulty was getting the software to keep track of each bird in a succession of images. More often than not, it was nearly impossible to distinguish one from the other and follow a single bird. Each seemed to be a dark speck amid hundreds or even thousands of other dark specks. The researchers, led by Andrea Cavagna and Irene Giardina, spent two years honing an algorithm to analyze the images and create reliable matches of individual birds across a series of photos. Ultimately, they found they could accurately examine and model flocks of about eight thousand starlings. Finally, there was a way to quantitatively see starling murmurations and begin teasing apart how the birds moved. In particular, they could compare individuals within a flock in search of basic rules that governed starlings' flight patterns.

What they found was that starlings avoided collisions by staying at least a wing's length away from one another and didn't stray far from the group. But how did they know when to turn left or right, dip or move skyward? The researchers landed on a novel rule of thumb that each flying starling takes its directional cue from its nearest seven neighbors. And those cues then ripple and multiply bird by bird through the flock, like a wave of information passed almost instantaneously through the group until all seem to be operating as one.

The study also provided a variation on a previous notion about collective animal behavior that assumed each individual responded only to neighbors within a fixed distance. The modeling of the Roman starlings found that the distance to the nearest neighbor didn't matter as much; each starling was reacting to seven neighbors no matter how far away they were. "In a packed flock, the seven neighbors you are interacting with are close to you, whereas in a loose, sparse flock they are more distant," Cavagna and Giardina wrote in an explanation of their work headlined "The Seventh Starling."[10]

That comparative distance between birds, what they called "topological" distance, turns out to be important because it allows a flock to stick more closely together and move more dynamically as a unit in the face of trouble. "Sticking together, keeping the cohesion of the group, is one of the key priorities of the flock, because isolated birds are dramatically more prone to predation than birds within the flock," the researchers said. "Not surprisingly, then, evolution has selected an interaction based on topological distance, which is more resilient to external attacks." But why seven? Cavagna and Giardina noted other experiments on birds that showed they could distinguish sets with different numbers of objects, as long as the numbers were less than seven. For example, they could differentiate between four and five but not between eight and nine. In the starling flocks, individuals may be able to see fifteen or twenty of their neighbors but not able to process all that complex information. "Seven may work as a cognitive limit for starlings: they simply cannot keep under control a larger number of neighbours," Cavagna and Giardina noted. It's possible that it's simply a coincidence that the limit of seven the researchers observed in starling flocks matches fairly closely with the captive bird experiments and Miller's Harvard study from the 1950s about "the magical number 7" among human test subjects. Or maybe there's something else at work. "Starling flocks provide yet another intriguing case to be added to this list," they said.[11]

BUT THERE WAS MORE TO BE GLEANED ABOUT STARLINGS from the skies above Rome. Starlings, perhaps twenty thousand strong, had spent the day outside Rome, feasting in the olive trees, and were now headed back into the city to settle for the night at their roost in Termini in the central part of the city. Before roosting, though, the starlings clustered in the sky in a massive flock, dancing in the air above the apartments, restaurants, and office buildings. Starlings had been roosting in the neighborhood for at least fifty years, generation after generation, and the murmurations were a common sight for the locals. Equally familiar with the routine were the opportunistic peregrine falcons that showed up in the unsettled air of midwinter dusk, typically just a few, with a mind to snatch a quick meal from this dark cloud of wheeling birds. And like the starlings at Palazzo Massimo, they were being watched.

On a nearby roof, cameras shooting high-definition video were trained on the drama overhead. The operators, part of a group of Italian researchers, had come more than an hour before dusk to set up, settle in, and try not to disrupt the proceedings. They wanted to document the hunting sequence and, more specifically, understand how the starlings communicated to each other about the impending threat. Just as specially selected priests had once tried to decipher the mysteries of the starling flocks in ancient Rome, these modern scientists working during the winter of 2006–7 were trying to unlock some of the secrets to the starlings' defense mechanisms in the face of one of the avian world's most efficient predators.

By then researchers had already documented a warn-the-others phenomenon in fish and some water bugs, called the Trafalgar Effect after the signals sent between ships in the British Navy to warn of a coming attack from the combined forces of the French and Spanish during the Napoleonic Wars. With the starlings in Rome, researchers were looking at how information about an impending attack—by a falcon, in this case—was transmitted through the flock at a rate faster than the predator could strike, just in time to avoid losing one of their own. This kind of collective behavior can be critical to the survival of prey species, and the starlings seemed to have it dialed in.

Over the course of several months, researchers filmed more than 300 hunting sequences above Termini and at another spot on the city's south side, where some sixty thousand starlings tended to congregate during the winter. Of those sequences, 210 produced an anti-predator reaction they called a "wave event," essentially a pulse of movement that traveled through the flock and away from the falcon. Through direct observation and analysis of the videos, the researchers could see that the individual starlings closest to the falcon turned away, and because they were flying so close together, that movement sent a ripple through the group, first in their neighbors and then in their neighbors' neighbors and so on. Sometimes the wave lasted a few seconds and other times it lasted up to twenty seconds, as it worked its way through thousands of birds swirling in the sky. The falcons were less successful when the waves were triggered, the scientists found. Not only did the starlings closest to the falcons live to see another day but the rest of the group did too. Call it the benefits of collective behavior.[12]

Years later researchers used the footage to identify and name other patterns that emerged when the starling flocks were attacked. Responses included "blackening," when starlings grouped tightly together; "flash expansion," when starlings moved out radially; a "split," when a flock briefly separated into subflocks; and "flock dilution," when the birds spread out, creating more space for individuals. Ultimately, they found six patterns of collective escape when starlings were under attack by falcons.[13]

MUCH OF THE WORK PROBING THE SECRETS OF ROME'S STAR-lings is part of a larger project, funded by the European Commission, called STARFLAG. It's a collaboration by European scientists from a range of disciplines—including biology, physics, and economics—looking at collective animal behavior and its potential for explaining behavior in other contexts. For example, the starlings in Rome, they said, may provide insight into how groups of humans behave, including in social situations; how we choose what to buy; or even what music we download.

Researchers at Princeton University and elsewhere have theorized about how to apply the principles of starling behaviors to robots, including taking their cues from seven neighbors. Group cohesion in a complex environment, such as when cooperative teams of robots work on a rescue mission in a dangerous place, may benefit from the starlings' delicate in-flight balance between individual decisions and those made to protect the overall flock. At Princeton, researchers Naomi Leonard and George Young said the insights about starlings helped create an algorithm allowing robots in a group to be more efficient and versatile, adjusting which other robots they interact with to adapt and enhance the performance of the group. "We are trying to draw inspiration from biology to understand what measures of animal group performance can help us decide what measures we should use when we design responsive behaviors for robots," Young said.[14]

But starling murmurations fascinate beyond computational models, mathematical equations, and obscure scientific journals. For most people, the starstruck observers among us, they occupy another realm, one dominated less by logic and more by mystery and delight.

New York–based photographer Richard Barnes captured images of Rome's starlings for a stunning black-and-white exhibit later titled *Murmur*. Rather than using a digital camera, he shot the images on film with a Hasselblad. "There was something about the way the grain of the film equated to the massive flocks of starlings in the sky that I really responded to when I started to process the negatives," he told *Slate*. "They swirl and weave and look strikingly like computer animation in the sky, creating pointillist abstractions one moment, only to disperse and then coalesce into what looks like a ball or a question mark the next."[15]

24

Built for Survival

EUGENE SCHIEFFELIN MAY NOT HAVE KNOWN IT, BUT THE starlings he let loose in Central Park long ago turned out to be one of nature's superstar survivors, royalty among birds when it came to fitting into a new place. As colonists go, few have been better in modern times. From its origins in Europe, *Sturnus vulgaris* now occupies every continent except Antarctica. Scientists in 2021 estimated there are around 1.3 billion European starlings scattered across the planet. Only house sparrows, with around 1.6 billion, are more numerous.[1]

To be sure, starlings are smart (their brains are larger than those of other, similar-sized birds), they breed rapidly, and they're bold, aggressive, and not easily rattled. They also possess an uncanny knack for finding food and making themselves at home in almost any environment. Plus they have a high tolerance for living cheek to cheek with humans in cities, sprawling suburbia, and rural outposts. "They are happy to scavenge scraps from parking lots, nest in derelict buildings, and roost in landfills. To us, those are lowlife traits—but that's pure anthropomorphization," ornithologist and author Noah Strycker told me. "They are remarkably successful in human-altered habitats."

Indeed, in recent decades, starlings have continued to make headlines in Boston, Indianapolis, rural Wyoming, Los Angeles, Pennsylvania, and all parts between. The local occupations follow a predictable pattern: a sudden arrival followed by a mess and plenty of noise, then curses and calls for help, teams dispatched to shoo them away, temporary relief, a brief reduction in numbers in targeted spots, an inevitable return, and sometimes a decision by the starlings to move along to somewhere else. For some people with an up-close view, part of the pattern includes a reluctant appreciation of the birds' substantial survival skills.

But as it turns out, part of the secret to their success happens beyond our observation. A groundbreaking genome sequencing study in 2021 revealed that while starlings in North America don't vary widely in their genetic makeup, there are signs of subtle differences, especially when it comes to adapting to variations in temperature and rainfall. In Arizona, for example, starlings' genomes showed strong evidence of adapting to hot, dry conditions. In the Pacific Northwest, the study revealed they'd become accustomed to cool, wet conditions. Notably, the changes happened extraordinarily fast. After starlings were introduced into Central Park, they fanned out across the continent, quickly adapting to regional conditions, with elegant tweaks in their genetic makeup that enabled the species to persist.

"The amazing thing about the evolutionary changes among starling populations since they were introduced in North America is that the changes happened in a span of just 130 years," said Natalie Hofmeister, who was the lead author of the study as a PhD candidate at the Cornell Lab of Ornithology. "For a long time we didn't think that was possible: that it took millions of years for genetic mutations to change a genome." The adaptations may not have been the result of a mutation but rather of an existing variation passed down through the generations of the first starlings that came to America. "A genetic variation that might not have been useful in one environment could turn out to be very beneficial in another," Hofmeister explained. "So, a variation related to temperature and rainfall that enhanced survival became more common in a new environment."[2]

Emerging science tells us that starlings possess other important evolutionary advantages. A 2023 study in *Science* showed that animals adept at complex vocal learning—like being able to learn and retain a large number of sounds—are also better problem solvers. That includes people, elephants, whales, and bats, but also songbirds like starlings.

Scientists at the Rockefeller University Field Research Center in New York's Hudson Valley spent years looking at vocal learning complexity in songbirds, including how many songs and calls are in a bird's repertoire, whether new songs can be learned over a lifetime, and the ability to mimic other species. Then they examined whether that learning was associated with other kinds of problem-solving. To do that, they caught hundreds of wild birds across more than twenty species that were living on the research center's twelve-hundred-acre reserve and put them through a series of tests. Among the tasks were removing a lid, piercing foil, and pulling a stick to retrieve a treat. Other tests looked at their ability to associate a certain color with a food reward and how quickly the birds adjusted when the color changed.[3]

Starlings, already known to be highly adept at vocalizations, were shown to be among the best when it came solving puzzles and working around obstacles to get a snack. Other vocal learners also did well, and there was a correlation to larger brain sizes relative to body size. The findings of the study suggested "vocal learning, problem solving, and brain size may have evolved in tandem," possibly as a way to help starlings and other gifted problem solvers survive in the wild. "Our next step is to look at the brains of the most complex species and try to understand why they are better at problem solving and vocal learning," said Jean-Nicolas Audet, the study's lead author. "We have a pretty good idea of where vocal learning happens in the brain, but it's not yet clear where the problem solving occurs."[4]

In another study several years earlier, scientists found that starling wings have become more rounded since the birds' arrival in North America more than a century ago. They examined more than three hundred dead starlings from museums collected since the 1890s and found a slight but measurable morphological change over time in wing shapes, speculating that the less-pointed wings may have helped starlings improve their foraging habits and better avoid predators. Studies have shown that birds with more rounded wings can take off at steeper angles and are better equipped to escape the talons of a hungry enemy. In the cutthroat game of survival in the wild, every advantage helps.[5]

Not long after my visit to Central Park to see where Schieffelin might have released his starlings, I talked with Julia Zichello, an evolutionary biologist who earned her PhD in primate genetics but has been studying starlings since around 2016. She also works at the American Museum of Natural History and is well familiar with the birds that Schieffelin had been so excited about when he learned they were nesting in the eaves of the building. Her interest was piqued when she spotted some "pretty black birds" on the lawn outside the museum's lab for exploring human origins and comparative genomics. She quickly got up to speed on starlings' origin story in the United States—their introduction and expansion, extraordinary adaptation skills, and subsequent reputation as irritating visitors—and like Hofmeister at Cornell, she wondered about their genetic homogeneity. Typically, genetic variation is a crucial driver of evolution. But was that the case with starlings on U.S. soil?

Working with fellow researchers and young students—and accepting delivery of garbage sacks full of dead starlings from government kill programs—Zichello had been trying to tease out the genetic differences between the starlings outside her window and those found elsewhere in the world, including in their native ranges in the United Kingdom and in other places where they were imported and released, like Australia and South Africa. Initial signs are that the starlings in the United States have much less genetic diversity than their cousins in their native lands, and yet they continue to thrive. The work has had her contemplating similarities between people and birds, especially when a small number of one species arrives in a new place and disperses in short order. There are all sorts of implications when it comes to genetic diversity, evolution, and adaptation.

"Human population history is also characterized by a relatively small population size expanding in a short period of time," Zichello said. "This has led to less genetic diversity in humans than in our close evolutionary relatives, the chimpanzees. So, I see conceptual parallels between starlings and humans, despite the two species being wildly different in certain ways."[6]

Outside her genetic research, Zichello still can't help but observe starlings everywhere she goes in New York City. She's seen them feasting on a pile of yellow rice on Columbus Avenue, chomping on a soft pretzel in Central Park, and digging into an apple pie in the Costco parking lot in Queens. It's hard not to respect their sheer ability to find food, modify their behavior for an urban environment, and survive. And at this point, it can be hard to imagine the city's avian life without the ever-present starlings hopping around, nesting in tree holes, and roosting high in the trees of Central Park. "To be honest, I really look at them with admiration," she told me. "But I've definitely explored the inner conflict. I know all about the terrible things they do, but do I still love them? Kind of, yes. I know I'm supposed to be a scientist and I'm supposed to be really cold, but I'm not."

BUT JUST AS WE'RE LEARNING MORE ABOUT WHAT HAS MADE starlings so successful, their global story has taken a darker turn. Although they've been wildly successful in their new homes, especially during the twentieth century, they've struggled in some of the places where they originated. Starling populations in Great Britain have dropped by more than half since 1964. In Finland, they fell 90 percent between 1970 and 1985. Denmark's numbers dropped by 60 percent between 1976 and 2015. Many of the declines have corresponded with changes in rural livestock operations in those areas, including steps to shield cow food from birds and moving some of the cattle indoors, reducing the amount of pasture. Other factors are still being investigated, including whether the urbanization of rural lands might be making life harder for starlings.

And as it turns out, starlings aren't what they used to be in North America either. After their population peaked at around two hundred million on the continent, today their ranks have probably fallen below one hundred million, perhaps around eighty-five million. Starlings' 50 percent decline in North America since 1970 is part of a staggering loss of birds from coast to coast in recent years. A shocking 2019 study found that the continent has lost nearly three billion birds over the past five decades or so. Warblers, larks, sparrows, and finches have been among the hardest hit, but by no means are they alone. In just two human generations, the skies have started to empty out with the loss of native and nonnative birds alike, including those thought to be common. The losses ripple through the natural world, where birds are relied on for controlling insects, pollinating plants, and dispersing seeds.[7]

The most likely explanation is changes in agricultural practices, said Kenneth Rosenberg, the study's lead author and a senior scientist at the Cornell Lab of Ornithology and American Bird Conservancy. "The intensification of agriculture is happening all over the world, [as is] increased use of pesticides, as well as the continued conversion of the remaining grass and pastureland—and even native prairie" to cropland, he said. Pollution and chemicals sprayed on crops can delay bird migrations, kill insects that birds eat, foul streams and rivers, and cause other problems. Loss of wild habitat, via pollution or plow, only makes things more difficult for most birds. Drastic human-caused changes to the climate, including the effects that cascade across ecosystems like waves crashing on a beach, add further insult.[8]

It seems likely that we've surpassed "peak starling" rates in North America. They might be hearty travelers, cognitive superheroes, and first-class colonizers, but even they aren't immune to the planet-altering forces that humans have unleashed upon the world. It's an unsettling truth that raises grave questions about what's in store for the rest of the natural world, not to mention us. "If our birds that are the most adapted to humans and cities and living among us are struggling," said Robyn Bailey, an avian biologist at the Cornell bird laboratory, "what does that mean?"[9]

Epilogue

ONE AFTERNOON LATE IN THE SUMMER OF 2022, MY FAMILY and I were driving south on Highway 91, just south of Springfield, Massachusetts, when a flock of starlings whooshed out of the trees crowding the Connecticut River. I did my best to keep an eye on the road but could hardly ignore the dark cloud of two hundred or so starlings writhing just above the horizon and fully ignorant of the traffic zipping by on the highway below them. They veered right, recovered, and then dipped again before changing their collective minds and pointing themselves back toward the dense trees at the river's edge. They were in and out of sight within just a few delightful seconds. Later that day we ran into a few more starlings in downtown Hartford, Connecticut. I spotted two perched on the window ledge of an office building and another resting atop a telephone wire. We'd come to drop my daughter off at college, but I was also on the lookout for the legendary hordes of starlings that once occupied the city and left so many of its residents in an uproar.

By the summer of 1914 millions of starlings were squatting in Hartford, roosting in the spires of St. Joseph's Cathedral and in the evenings taking refuge in the trees of the city's fanciest boulevard, called Washington Street. Residents of this tree-lined "street of governors," famed for its mansions that had long housed the city's elite political families, didn't remain quiet in the face of such an invading force. The scare tactics that city crews came up with included a combination of rockets fired directly at the birds and teddy bears hanging from the branches. In March 1931 so many starlings had crammed into the tower of the old statehouse building downtown that it stopped the clock that "thousands of people depend[ed] on" to set their watches and get the time of day, according to one report. It happened again at the end of 1938, when hundreds of starlings squeezed themselves onto the four-faced clock in the tower. "Time actually stood still in Hartford," the article quipped.[1]

While city officials contended with the nuisance, a biologist at nearby Trinity College was running a series of experiments on caged starlings, trying to better understand how the heat and intensity of daylight influences breeding cycles. This was part of a larger endeavor to understand similar phenomena in other species, including people.

When I arrived in Hartford nearly a century later, the starlings were scarce. I asked around at the local Audubon chapter and was told that fifty thousand or more starlings roosted near Union Station a few years earlier, but they had mostly vanished, dispatched by some unseen hand. "The starlings in Hartford are but a whisper of what they once were," I was told by a local bird expert.

The neighborhood they'd once seized had changed too. What was once a tony stretch of mansions buckling under the occupation of starlings was rougher now, dotted with boarded-up windows, overgrown yards, and people on the sidewalks who seemed lost as traffic whizzed past. Walking along, I paused to imagine a shady street a century earlier awash in the cacophony of starlings' chaotic calls—what was the sound of a million stubborn birds singing?—but struggled to fall under the spell of a noise I could scarcely call up. They felt like phantom starlings now, or perhaps simply travelers already dispersed and departed.

IT WASN'T SO LONG AGO THAT A GROUP OF INTERNATIONAL scientists named starlings as one of the top hundred "invasive alien species," those considered the most dangerous and damaging for their ability to "establish, thrive and dominate in new places." They shared the list with some of the most infamous invasive plants, animals, and fungi, including brown tree snakes, feral pigs, avian malaria, mosquito fish, leafy spurge, and crazy ants.[2]

Separately, a 2023 report from the United Nations' Intergovernmental Platform on Biodiversity and Ecosystem Services said that the "severe global threat posed by invasive alien species is underappreciated, underestimated, and often unacknowledged." Some thirty-seven thousand species have been transported to new regions over the past few centuries, sometimes intentionally and sometimes not. Of those, the report said, thirty-five hundred are harmful and are jeopardizing native ecosystems, spreading disease, and threatening crops and infrastructure. The cost for attempting to control them now tops $420 billion per year.[3]

We've seen in recent decades what invasive species can do, including kudzu vines that choke plants in the South, zebra mussels that clog water intake pipes in the Great Lakes, brown tree snakes that make a meal out of native birds in Guam. Some 60 percent of wildlife extinctions can be partially attributed to invasive species. I haven't found any reports that starlings in North America are threatening to drive any species into oblivion (at least, not yet), but the other damage and distress they've provoked have been pretty clear over the past century or more. So despite declines in overall numbers, there's still no shortage of animosity toward starlings today. Most of my bird-loving friends and colleagues turned up their noses at my interest in them, putting them into the category of "trash birds" that deserve little of our affection or interest. One invited me over to watch him capture and gas a few starlings that had taken over the cavities of a backyard tree. Another, when I asked if she ever trained her binoculars on starlings when she was out bird-watching, just shook her head and made a face as if I'd asked if she enjoyed smelling dirty socks.

Most of the empathy in the equation seems to flow toward people, not starlings. "You are not alone," one bird-shooing business assures visitors on its starling web page. Among the options now available for those trying to get rid of starlings are pole-mounted, high-tech noise devices that play a random selection of bird calls, including distress calls and predators' screeches, designed to clear birds from farms, marinas, buildings, and vineyards. There are electric shock machines and special repellent gels designed to affect birds' sight, smell, and touch. Some might go with slick slime to keep birds off ledges and nets to keep them off crops. People still put out fake owls and hoist oversize balloons designed to look like giant predatory eyeballs. Lasers, water hoses, and explosives are still used. In some places, you can even hire a specially trained hawk to come out and scare away your unwanted starlings.[4]

Research on starlings continues apace, too, a sign that the puzzle remains largely unsolved. One recent study, carried out at a rural NASA research site in northern Ohio, tested whether starlings might be scared off their nests if they spotted a snake. They didn't use real snakes, of course, but placed spring-loaded fake snakes—constructed from black cords and a conical "head" with fake eyes and placed in ready-to-strike form—inside a series of nesting boxes attached to utility poles. Other boxes had fake snakes that didn't move. The hope was that starlings would be frightened and either delay nesting or move somewhere else. No such luck: starlings moved into 76 percent of the boxes and went about their business. Fake snakes, at least like the ones used in the experiment, were a bust.[5]

In Nebraska, a researcher at the U.S. Department of Agriculture has been working for more than a decade on a better trap for starlings. He got involved after thousands of starlings set up shop in 2004–5 in downtown Omaha, where deploying DRC-1339 proved difficult because of the urban environment. He spent several years tweaking his design, and in the winter of 2011 he caught about twenty-five thousand in his experimental wire cages. The birds were then gathered up and gassed. His four-chamber contraption, still a work in progress, reduced Omaha's starling population by 90 percent, he said. His 2021 paper, though, stopped short of claiming victory. He titled it "One Step Closer to a Better Starling Trap."[6]

At the College of William and Mary in Virginia, researchers used seventy wild-caught starlings to test out a "sonic net," a series of ultrasonic beams of sound trained on food sources in an enclosed aviary. The array of highly directional sounds was of a similar frequency to starling vocalization, intended to disrupt the communication among starlings and prompt them to leave the food alone and go elsewhere. The number of starlings at the food source decreased by about 46 percent, the researchers said, enough to declare it "a notable achievement." More testing would be needed, though, including experiments in the field.[7]

In some respects, we've come a long way from the days of Bird Men chasing birds around building rooftops and people blasting vinyl records. On the other hand, it sometimes feels as if we're as flummoxed as ever, perpetually casting about for solutions to a problem we know will never really go away.

I'll admit to mixed feelings about *Sturnus vulgaris*. Like so many things when it comes to nature, it's complicated, but I can't help but sometimes root for the underdog, especially species that seem to be under relentless pursuit by people. Where so many see an obnoxious pest, I find myself looking closer at starlings' iridescent feathers and the speckles that call to mind an abundance of stars in the night sky. I love that they walk with one foot in front of the other, looking more like a wobbly toddler than an elegant and distinguished bird. To me, their murmurations at dusk are works of art, even more precious because their giant collective shapes are so liquid and fleeting. The fact that they can mimic the sounds of so many animate and inanimate objects—the tweet of a key fob and the notes of a Mozart composition—never gets old. Even the cacophony of thousands in a roost is something to admire, sheets of furious noise like something from John Coltrane's last years that turned so many people off, blasts of sonic eruptions that insist on being heard. So much of the human world has become about control and predictability. It's good to be reminded that nature, too, will have its say. And shouldn't it?

I know there has been a cost from starlings but we can't disown the fact that we purposely brought them to North America. They were captured in their homeland, transported across the ocean, and turned loose on foreign soil. No starling asked to come here, as far as I know, so it seems unfair to view them with contempt for being themselves in a land that was never their home. What were they supposed to do? Our culpability can be expanded for making the continent more hospitable to starlings by mowing down forests, building cities, and plowing up wild land. We assembled airports, opened massive cattle operations, seeded orchards, erected intricate buildings, and planted towering neighborhood shade trees—each another enticement for these birds determined to evolve right alongside us. Predictably, we grew annoyed and responded with poisons, explosives, goo, noise, and all manner of deterrents. Starlings answered with perhaps their most potent trait: a stubborn iron will to simply carry on, year after year, decade upon decade.

But they're diminished for now, a shadow of their heyday in North America, victims of the human condition, which seems to bring low so many other wildlife species. The startling loss of birds in the past fifty years is but one small chapter in the larger story being written around us, spurred by habitat we've destroyed, air and water we've contaminated, nonnative species we've introduced, a climate we've altered with little regard for the long-term consequences. Scientists predict one million species could vanish in the coming decades as part of the most significant wildlife extinction event that our species has ever experienced. In the face of such profound losses of animals and plants, I have to wonder if it's time to rethink our approach to European starlings in North America, even just a little. With so many birds already gone and many more faced with the possibility of extinction, is it really necessary to harass and kill so many of those that remain, even if they're not originally from these parts?

Of course, first and foremost we must protect native species and do whatever we can to shield them from extinction—and sometimes that means killing invasive species to protect native ones. But things become fuzzier after that. At what point do visitors become residents? And precisely how long is it before an invited guest—left with no option but to stay indefinitely—is welcomed as a permanent neighbor?

I won't pretend that tolerance for starlings or any other invasive species in the United States makes sense in the cold light of ecological logic. I certainly don't want to live in a world where local native species, especially those adapted to live in just a few places, are ousted by generalist species from afar, including those whose spread is facilitated by our own foolish behavior. Homogeneity at the cost of biodiversity would rob us in profound and innumerable ways. So I feel no allegiance to Burmese pythons killing endangered wood storks and wreaking havoc in Florida or to mosquitoes wiping out rare native birds in Hawaii. Good riddance to them and those like them. And starlings? Sound reasoning would point toward doing everything we can to rid ourselves of them and to see them as they are: a species that simply doesn't belong here and whose presence only diminishes the astonishing suite of animals and plants that evolved on this continent naturally over the eons.

Still, there's part of me that feels obligated to join the call taken up by Rachel Carson, Edmund Selous, and others to be a few degrees more tolerant toward starlings. They're among us now, and they're not going away. Rather than reaching for a gun or explosives, perhaps we should occasionally pick up a pair of binoculars and look at these peculiar and intelligent birds in a new way. Maybe in a moment of suspended disgust, there's a chance to reflect on one of nature's flying wonders, however out of place. Difficult choices abound when it comes to balancing the interests of starlings and, say, other native birds or human enterprises that are dirtied or worse by their existence. Harsh and lethal measures will still be needed at times, I'm sure, but it seems worth considering whether a slight shift in attitude and approach might benefit all involved. After all, we're still getting to know these strangers.

As it turns out, while so many people were wringing their hands about the starling problem, more than a few kept them as pets. Years ago, writer Lyanda Lynn Haupt adopted a five-day-old starling, named it Carmen, and kept it as a pet. Her book *Mozart's Starling* chronicles their time together, including Carmen's delightful habit of sitting on her shoulder while she wrote. Online, there are forums devoted to all manner of starling care, along with the antics of those allowed into homes. I was struck by a minidrama that unfolded in 2019, when a one-legged pet starling named Gordo went missing in Galloway, Ohio. The family had adopted him after he showed up injured in the family's yard and the local wildlife rehab place declined to take him in. Soon Gordo was talking ("Whatcha doin'?"), whistling the tune of "Shave and a Haircut," and mimicking their cell phone ring. When he went missing, the owners put up lost-bird flyers around town that include a photo of Gordo. "I just want him back," the owner told the local newspaper. "I guess that sounds selfish, because he's a wild bird, but he grew up with us."[8]

IN THE FALL OF 1925 A MOVING ESSAY BY FRANK M. CHAPMAN, one of the most famed and respected ornithologists of his time, appeared in *Natural History*, the bimonthly journal of the American Museum of Natural History. In it, he seemed to be sorting out his feelings about starlings in plain view of the reader. Chapman, sixty-one, had spent much of his professional and personal life studying and thinking about birds. He'd already published several books, been elected to the National Academy of Sciences, worked as the first curator of birds at the Museum of Natural History, spoken up for the plight of the nation's declining native birds, founded the annual Christmas bird count, and started the influential *Bird-Lore* magazine. As the dean of American ornithologists, he could have trained his attention on any number of bird issues of his time, but as he had done often over the years, Chapman felt compelled to examine the starling question. A central conflict clearly weighed on him: How was it that a thing of such beauty and fascination was also so repulsive to so many in its new home?

He laid the blame for the conundrum at the feet of Eugene Schieffelin, who had thought the deficiency of birds in America might be filled by starlings, house sparrows, and others from across the Atlantic. "Inspired by the highest motives, he might, under proper direction, have become the father of bird conservation in America and have saved for us species we shall never see in life again," Chapman wrote. "But like many another pioneer reformer, he blazed a false trail."[9]

It wasn't difficult to predict that Schieffelin's high aspirations for his starlings would turn bitter when thousands of them came to the city in the late summer and early fall. "There are times when the starling makes such demands upon our hospitality that even its friends resent its presence," Chapman said. "To the ornithologist such gatherings possess great interest, but the tired business man, who has come home to sleep, not to study avian communism, sees in them only a source of noise and filth, and spares no effort to induce the birds to move elsewhere. Frequently, however, he finds it difficult to convince them that they are not welcome."[10]

But it was more than just the roosting habits. Starlings arrived as foreigners, inevitably disturbing the established order of things and finding themselves in a diminished standing among neighbors. The baggage of being an alien species followed them wherever they went, at least among those paying attention. If a blur of birds in the sky turned out to be a flock of red-winged blackbirds in March, Chapman noted, it was celebrated as a tender sign of spring. "But if the hurrying smudge becomes a passing troupe of starlings, we regard it with disappointment or indifference," he said. Similarly, if keen listeners thought they were hearing the voice of a wood-pewee in the summer air, but it turned out to be an imitation by a starling, it was considered "a mimetic travesty," Chapman said. "But, after all, it is only the chosen few who cherish these intimate associations with nature that resent the starling's violation of them."[11]

He also cautioned against burdening starlings with too much of our own human context. Considered simply as a bird, starlings have much to recommend them, including their summer dress of black with that distinctive iridescent sheen, not to mention their winter outfit with playful polka dots. Their song, too, deserved consideration. Yes, they could mimic others, but they also featured a durable note worth listening to, "a long drawn, cheerful whistle, human-like in quality and easily imitated."[12]

Chapman saved his greatest admiration for starlings' aerial acrobatics, sky scenes where perhaps ten thousand birds, "animated by one impulse," changed in a split second from a globe to a dumbbell to a dusky snake and back to a ball again—all performed with precision, with no clear leader and no apparent word of command. "It is," he said, "a dance in the clouds to the music of the winds—a pure expression of a *joie de vivre*, which raises the industrious plodder of our lawns to an ethereal realm where nationalities are unknown and the glorious heritage of flight is the universal emblem of bird life."[13]

FOR ALL THE BIRDS THAT WHEEL AND WING THROUGH EMILY Dickinson's poems, it's doubtful she ever saw a starling in the wild with her own eyes. By the time she died in 1886, starlings had been released just a few times in North America, in places like West Chester, Pennsylvania, and Cincinnati, Ohio. The closest to her home in Amherst, Massachusetts, may have been the 1877 release in New York City's Central Park. None of those releases amounted to much, and starlings didn't really become established until Eugene Schieffelin's round of releases in the late 1880s and early 1890s.

Still, while I was on the trail of starlings, I came back again and again to her lovely if cryptic poem "Hope." Its three stanzas go like this:

Hope is the thing with feathers
That perches in the soul,
And sings the tune without the words,
And never stops at all,
And sweetest in the gale is heard;
And sore must be the storm
That could abash the little bird
That kept so many warm.
I've heard it in the chillest land,
And on the strangest sea;
Yet, never, in extremity,
It asked a crumb of me.[14]

Although the poem was included in a handmade collection of Dickinson's poems in the 1860s, it was first printed and commercially sold in 1891. I wondered often if Eugene Schieffelin had ever read it. His final release of starlings into Central Park was in the spring of that same year. Maybe he hadn't seen Dickinson's poem by then, but I can imagine him, sometime later, coming across its brief stanzas and feeling some deep kinship between her words and his deeds. In its purest form, Schieffelin's releases were an act of optimism and faith. Profoundly misguided, perhaps, but he intended for his birds to make the world a better, more beautiful place. After all, who could not wish for more bird songs? How could the skies not benefit from a few more wings fluttering wildly in the air?

And if hope is the thing with feathers, why not fill this confounding, uncertain existence with feathers upon feathers?

Acknowledgments

IF STARLINGS HAVE DONE NOTHING ELSE, THEY'VE STIRRED up emotions and sent many a person—either in delight or in frustration—to their pens, their calculators, their cameras, or their lists of novel ideas for shooing them away. Few birds have elicited such a potent response, and it's gratifying that in our modern, mechanized world, increasingly isolated from nature, a wild animal can still cause a fuss.

I'm grateful to all the ornithologists, historians, bird-watchers, poets, bureaucrats, befuddled citizens, writers, photographers, painters, and others who have tried to understand starlings and their occupation of North America. It's a complicated story that has unfolded over more than a century in North America, with still more chapters to be written. Thanks to all those who came before me to tell a portion of the story, and deep appreciation to those who talked with me, answered questions, passed along hard-to-find documents, and opened new avenues for exploration. My journalistic history of starlings is built on the knowledge that many others have developed over years and years in the field, behind binoculars, in the lab, and often in obscurity. I've done my best to capture much of their work here and do justice not only to these fascinating birds but also to the equally fascinating human response to their presence. Any mistakes are my own.

Thanks, too, to all the fine and dedicated folks at Bison Books and the University of Nebraska Press who have supported this project and continue to make this world better by fostering a love for books, stories, and words.

As always, a note of humble appreciation for my own little flock, especially Karen (expert reader and sharp editor) and Birdie, both of whom have endured my quirky obsession with these often-unloved birds as if it's all quite normal. Fly on.

Notes

1. The Bird Man Cometh

1."Otto and the Night Visitors," *Sports Illustrated*, August 31, 1959.

2.McCandlish Phillips, "Bird Man Raises His Flaps Again," *New York Times*, August 22, 1959.

3."Otto and the Night Visitors."

4."Blackbirds and Starlings Cloud Streets of Mount Vernon at Dawn; 100 Found Dead," *New York Times*, October 29, 1929; Estelle F. Kleiger, "Best in Mount Vernon," *New York Times*, September 16, 1990.

5.John C. Devlin, "50,000 to 100,000 Starlings Commute Daily to 125th St.," *New York Times*, March 16, 1959.

6.E. R. Kalmbach and I. N. Gabrielson, *Economic Value of the Starling in the United States*, U.S. Department of Agriculture Bulletin No. 868, January 10, 1921, 46.

7.Rachel L. Carson, "How About Citizen Papers for the Starling?" *Nature*, June–July 1939, 318–19; Robert Cantwell, "A Plague of Starlings," *Sports Illustrated*, September 9, 1974.

8."Mount Vernon Starlings Flee Din of Traveling Bird-Chaser," *New York Times*, August 21, 1959.

9.Phillips, "Bird Man Raises."

10."Otto and the Night Visitors."

11."Starlings Confer as Bird Man Rests," *New York Times*, August 29, 1959.

12."Starlings Fly Back to Mount Vernon," *New York Times*, July 19, 1960.

2. Mr. Schieffelin's Birds

1.*One Hundred Years of Business Life, 1794–1894* (New York: W. H. Schieffelin, 1894).

2.Walter Spooner, ed., *Historic Families of America*, vol. 3 (New York: Historic Families Publishing Association, 1907).

3. *New York Genealogical and Biographical Record* 37 (1906): 317.

4. "Snap Shots," *Forest and Stream* 32, no. 14 (April 25, 1889): 273.

5. Board of the Department of Public Parks, *Report of the Central Park Menagerie*, Document No. 115, February 19, 1890, 9.

6. "Summer Plans for the Zoo," *New York Evening World*, April 9, 1890.

7. Edwin Way Teale, "In Defense of the Pesky Starling," *Coronet* (November 1947): 96.

8. Lauren Fugate and John MacNeill Miller, "Shakespeare's Starlings: Literary History and the Fiction of Invasiveness," *Environment Humanities* 13, no. 2 (November 2021): 301–22.

9. Fugate and Miller, "Shakespeare's Starlings," 311–12.

10. Fugate and Miller, "Shakespeare's Starlings," 312.

11. Cantwell, "Plague of Starlings."

3. A Frenzy of Introductions

1. Christopher Lever, *They Dined on Eland: The Story of Acclimatisation Societies* (London: Quiller, 1992), 3–4.

2. Thomas R. Dunlap, "Remaking the Land: The Acclimatization Movement and Anglo Ideas of Nature," *Journal of World History* 8, no. 2 (Fall 1997): 303–19.

3. "New York Board of Aldermen, 1871," in *Documents of the Board of Aldermen and Board of Assistants of the City of New York* (New York: New York Printing Company, 1871), 258.

4. "Acclimating Foreign Birds," *Forest and Stream* 8, no. 17 (May 30, 1877), 262.

5. "American Acclimatization Society," *New York Times*, November 15, 1877.

6. "The Starling in Central Park," *Forest and Stream* 8, no. 19 (June 14, 1877), 307.

7. "Starling in Central Park."

8. "The Birds of Central Park," *New York Daily Tribune*, June 22, 1890.

9. Frank M. Chapman, "The European Starling as an American Citizen," *Natural History* 25, no. 5 (September–October 1925): 479–85.

10. Cantwell, "Plague of Starlings."

11. Frank M. Chapman, *Handbook of Birds of Eastern North America* (New York: D. Appleton, 1895), 259.

12. Edmund Selous, *Bird Glimpses* (London: George Allen, 1905), 145.

13. Nicholas Lund, "Birdist Rule #72: It's Okay to Dislike Some Birds," The Birdist's Rule of Birding, *Audubon*, March 9, 2016.

4. The Sparrows

1. Peter Coates, "The Avian Conquest of a Continent," in *American Perceptions of Immigrant and Invasive Species* (Berkeley: University of California Press, 2006), 35.

2."Pittsburg's [sic] Exciting Battles in Mid Air," *Los Angeles Herald*, November 14, 1897.

3."Skinner's Sparrow: An Exotic Bird," *Spoonbill* 27, no. 26 (October 1978): 19.

4.Walter B. Barrow, *The English Sparrow (Passer domesticus) in North America, Especially in Its Relations to Agriculture*, U.S. Department of Agriculture Bulletin No. 1 (Washington DC: Government Printing Office, 1889), 17.

5."Skinner's Sparrow."

6."The English Sparrow," *St. Paul Daily Globe*, September 11, 1882.

7.*Annals of the Lyceum of Natural History of New York City*, vol. 8 (New York: William Wood, 1867), 287.

8."Our Feathered Friends," *New York Times*, November 22, 1868.

9.Coates, "Avian Conquest."

10.Barrow, *English Sparrow*, 22.

11."Our Sparrows," *New York Times*, November 20, 1870.

12."Skinner's Sparrow"; "The Sparrow," *New York Times*, October 3, 1884.

13."Pittsburg's Exciting Battles."

14."Natural History," *Forest and Stream* 17, no. 3 (August 18, 1881): 46.

15."Natural History," 46.

16."The Sparrow."

17.T. S. Palmer, "A Review of Economic Ornithology in the United States," *Yearbook of the U.S. Department of Agriculture* (Washington DC: U.S. Government Printing Office, 1899), 280; T. S. Palmer, "The Danger of Introducing Noxious Animals and Birds," *Field and Stream*, June 10, 1899.

18.Barrow, *English Sparrow*, 13.

19.Barrow, *English Sparrow*, 22.

20."Skinner's Sparrow."

21."The Sparrow."

22."Pittsburg's Exciting Battles."

23."Pittsburg's Exciting Battles."

24.William Leon Dawson, *The Birds of Ohio* (Columbus OH: Wheaton, 1903), 40.

5. Across the Sea in Cages

1."A Seminary for Teaching Birds How to Sing," *Scientific American* 79, no. 2 (July 9, 1898): 23.

2."Importing Song Birds for Our Woods," *Scientific American* 79, no. 16 (October 15, 1898): 243.

3.E. Kimbark MacColl, *The Shaping of a City: Business and Politics in Portland, Oregon, 1885–1915* (Portland OR: Georgian, 1976), 194.

4."Sweet Singers of Oregon," *Corvallis Gazette*, January 18, 1906.

5."Importing Song Birds," 243.

6."Sweet Singers of Oregon."

7."Music from Europe," *Seattle Post-Intelligencer*, January 15, 1891.

8."Importing Song Birds," 243.

9. "The European Starling in Oregon," *Oregon Birds* 19, no. 4 (Winter 1993): 93.
10. Erkenbrecker Memorial Committee, *Andrew Erkenbrecker* (Cincinnati: Erkenbrecker Memorial Committee, 1908).
11. Greg Hand, "The Unintended Consequences of Mr. Andrew Erkenbrecker," *Cincinnati*, March 3, 2016.
12. Lever, *They Dined on Eland*, 184.
13. Erkenbrecker Memorial Committee, *Andrew Erkenbrecker*.
14. Lever, *They Dined on Eland*, 185–86.
15. *Journal of the Cincinnati Society of Natural History* 4, no. 4 (December 1881): 343.
16. Eugene Schieffelin to C. F. Pfluger, May 23, 1893, Oregon Historical Society Research Library, Portland.
17. Schieffelin to Pfluger, August 18, 1894.
18. Schieffelin to Pfluger, May 23, 1893.

6. Lessons from Down Under

1. T. S. Palmer, "The Danger of Introducing Noxious Animals and Birds," U.S. Department of Agriculture (USDA), 1898, 93.
2. "Millions of Rabbits," *Olneyville Times*, July 12, 1889.
3. Palmer, "Danger of Introducing," USDA, 1898, 93.
4. Dunlap, "Remaking the Land," 315.
5. Walter W. Froggatt, "The Starling: A Study in Agricultural Zoology," *Agriculture Gazette of New South Wales* (C. Potter, July 1912), 610.
6. Palmer, "Danger of Introducing," USDA, 1898, 102–3.
7. Froggatt, "Starling," 613.
8. Froggatt, "Starling," 615.
9. Palmer, "Danger of Introducing," USDA, 1898, 108.
10. Palmer, "Danger of Introducing," USDA, 1898, 98, 109.
11. Palmer, "Danger of Introducing," *Field and Stream*, 1899.
12. 33 Cong. Rec. H4871 (April 30, 1900).

7. Occupation

1. Frank M. Chapman, "Our New Bird Citizen," *Country Life in America* 12, no. 5 (September 1907): 551.
2. Herbert K. Job, "Danger from the Starlings," *Outing* (Winter 1910): 146.
3. Job, "Danger from the Starlings," 147–53.
4. Edward Howe Forbush, *The Starling*, Massachusetts State Board of Agriculture Circular No. 45 (Boston: Wright & Potter, 1916), 12–19.
5. "Montclair Killing Starlings," *New York Times*, August 18, 1911.
6. "Bird Slaughter Denounced," *New York Times*, August 20, 1911.
7. "Man Who Slew Birds at Officials' Request, Jailed," *Newark Evening Star*, August 31, 1911; "Montclair to Fight Jailing of Stevens for Shooting Birds," *Newark Evening Star*, September 1, 1911.
8. "Finds No Bird Slaughter," *New York Times*, August 27, 1911; "Montclair May Kill Birds," *New York Times*, August 8, 1912.

9. "War on Army of Birds with Noise," *Newark Evening Star*, August 6, 1912; "Montclair's Avian Pests," *New York Times*, August 11, 1912.

10. "War on Army of Birds."

11. "War on English Starlings," *Cape County Herald*, May 10, 1912; "Stamford Gunner Fined $14.69 for Killing Starling," *Bridgeport Evening Farmer*, October 7, 1911.

12. "City at War on Starlings," *New York Times*, July 30, 1914; "The Starling at Springfield, Mass.," *Auk* 29 (1912): 243.

13. "Memories of Shad Seasonable during Month of March," *Bridgeport Evening Farmer*, March 10, 1914.

14. "English Starling Killed Near Savannah Last Week," *Americus Time-Recorder*, November 27, 1917.

15. "Various Matters," *Norwich Bulletin*, January 1, 1920.

16. "Starlings Fight Songbirds; Man Shoots Them; Arrested," *New York Times*, June 1, 1924.

17. "Eugene Schieffelin Dead," *New York Times*, August 16, 1906.

8. European Origins

1. C. G. Abbott, "European Birds in America," *Bird-Lore* 5, no. 5 (September–October 1903): 163.

2. Forbush, *Starling*.

3. Diana Wells, *100 Birds and How They Got Their Names* (Chapel Hill NC: Algonquin Books, 2001), 235.

4. Bob Sundstrom and Mary McCann, "Starlings and Roman Divination," *Bird Note* podcast, National Audubon Society, September 20, 2020, https://www.birdnote.org/listen/shows/starlings-and-roman-divination.

5. Job, "Danger from the Starlings," 148.

6. Forbush, *Starling*, 5–6.

7. Margaret Morse Nice, "An Appreciation of Edmund Selous," *Bird-Banding* 6 (July 1935): 95.

8. Edmund Selous, *Bird Life Glimpses* (London: George Allen, 1905), v.

9. Selous, *Bird Life Glimpses*, 141–42.

10. Selous, *Bird Life Glimpses*, 137–38.

11. Selous, *Bird Life Glimpses*, 138–40.

12. Selous, *Bird Life Glimpses*, 141.

13. Selous, *Bird Life Glimpses*, 142.

14. Selous, *Bird Life Glimpses*, 151.

15. Selous, *Bird Life Glimpses*, 143.

16. Selous, *Bird Life Glimpses*, 157–59.

9. The Skies Transformed

1. Nancy Plain, *This Strange Wilderness: The Life and Art of John James Audubon* (Lincoln: University of Nebraska Press, 2015), 24–26.

2. S. G. Morley, "A Note on Passenger Pigeons in the Nineteenth Century," *Condor* 37, no. 2 (March–April 1935), 87.

3. John C. Phillips, *Wild Birds Introduced or Transplanted in North America*, Technical Bulletin No. 61, U.S. Department of Agriculture, 1928, 1.

4. Phillips, *Wild Birds Introduced*, 1.

5. Phillips, *Wild Birds Introduced*, 5.

6. Frank Bolles, "Bird Traits," *New England Magazine* 7, no. 1 (September 1892): 96.

7. Coates, "Avian Conquest," 35.

10. Appetites

1. Jack F. Welch, "In Memoriam: Edwin Richard Kalmbach," *Auk* 90 (April 1973): 364–74.

2. Palmer, "Review of Economic Ornithology," 259–60.

3. "Official Champions English Sparrow," *Hawaiian Star*, October 25, 1911.

4. Kalmbach and Gabrielson, *Economic Value*, 27.

5. "Starling Called Gardener's Friend," *Washington Times*, March 28, 1920.

6. Kalmbach and Gabrielson, *Economic Value*, 59.

7. Christopher Feare, *The Starling* (Oxford: Oxford University Press, 1984), 18–19.

8. Henry Mousley, "Further Notes and Observations on the Birds of Hatley, Stanstead County, Quebec, 1919–1923," *Auk* 41, no. 4 (October 1, 1924): 584.

9. "European Starling Spreading Westward," U.S. Department of Agriculture, press release, March 13, 1925.

10. Marcia Brownell Bready, *The European Starling on His Westward Way* (Albany NY: Knickerbocker Press, 1929), 27.

11. "Warden Starts War on Predatory Birds," *Washington (DC) Evening Star*, June 8, 1930.

12. Associated Press, "European Starling Threatening Crops," *Oregon Statesman*, October 6, 1929.

13. May Thacher Cooke, *The Spread of the European Starling in North America*, Circular No. 40, U.S. Department of Agriculture, 1928.

14. "Starling Invasion a Police Problem," *New York Times*, January 6, 1933.

11. Sing a Song of Starlings

1. Bready, *European Starling*, 66.

2. Bready, *European Starling*, 63.

3. Alice E. Ball, *Bird Biographies: A Guide-Book for Beginners* (New York: Dodd, Mead, 1923), 80.

4. Bready, *European Starling*, 71.

5. Charles W. Townsend, "Mimicry of Voice in Birds," *Auk* 41, no. 4 (October 1924): 545.

6. Winsor M. Tyler, "The Starling as a Mimic," *Auk* 30, no. 3 (July–September 1933): 363.

7. Henry L. Saxby, *The Birds of Shetland, with Observations on Their Habis, Migration, and Occasional Appearance* (Edinburgh: MacLachan & Stewart, 1874), 117.

8. Townsend, "Mimicry of Voice in Birds," 544.

9. Quoted in Marcel Eens, "Understanding the Complex Song of the European Starling," in *Advances in the Study of Behavior*, vol. 26, ed. Peter J. B. Slater, Jay S. Rosenblatt, Charles T. Snowdon, and Manfred Milinski (Cambridge MA: Academic Press, 1997), 358.

10. Froggatt, "Starling."

11. A. M. Hindmarsh, "Vocal Mimicry in Starlings," *Behaviour* 90 (1984): 303.

12. Lyanda Lynn Haupt, *Mozart's Starling* (New York: Little, Brown Spark, 2017), 31.

13. Haupt, *Mozart's Starling*, 245.

14. Townsend, "Mimicry of Voice in Birds," 544.

15. Eens, "Understanding the Complex Song," 362.

16. Quoted in Townsend, "Mimicry of Voice in Birds," 543–44.

17. Meredith J. West and Andrew P. King, "Mozart's Starling," *American Scientist* 78, no. 2 (March–April 1990): 108.

18. West and King, "Mozart's Starling," 108.

19. West and King, "Mozart's Starling," 109.

20. Linda Weiford, "Like a Dog, like a Frog, like a . . . Starling?," *Washington State University News*, May 2, 2016.

12. Under Siege

1. Myron H. Swenk, "A Second Record of the European Starling in Nebraska," *Nebraska Bird Review* 1, no. 1 (January 1933): 15.

2. "New Pest in Form of Bird, Is Working Westward," *Frontier* (O'Neill City NE), February 22, 1934.

3. "General Notes," *Nebraska Bird Review* 6, no. 1 (January–June 1938): 4.

4. "General Notes," 6.

5. Arthur L. Goodrich Jr., "Starling Attacks upon Warble Infested Cattle in the Great Plains Area," *Journal of the Kansas Entomological Society* 13, no. 2 (April 1940): 35.

6. Goodrich, "Starling Attacks," 36.

7. Goodrich, "Starling Attacks," 36.

8. Leonard Wing, "Spread of the Starling and English Sparrow," *Auk* 60, no. 1 (January 1943): 74–87.

9. E. R. Kalmbach, "Winter Starling Roosts of Washington," *Wilson Bulletin* 44, no. 2 (June 1932): 65.

10. D. C. Peattie, "Starlings," *Washington (DC) Evening Star*, February 2, 1926.

11. Kalmbach, "Winter Starling Roosts," 68.

12. Kalmbach, "Winter Starling Roosts," 72.

13. "Starlings Become Cemetery Pests," *Washington (DC) Evening Star*, August 26, 1930.

14. "The Starlings: They Help Us but Often They Make Trouble," *New York Times*, December 20, 1931.

15. "Archives Building Badly Damaged," *Washington (DC) Evening Star*, March 24, 1935.

16. "Advises Boiling of Starlings' Eggs to Reduce Pests," *Washington (DC) Evening Star*, January 20, 1935.

17. "Advises Boiling."

18. "Advises Boiling."

19. "Advises Boiling."

13. In Defense of Starlings

1. Rachel Carson, "How about Citizenship Papers for the Starling?," *Nature Magazine* (June–July 1939): 318.

2. Carson, "How about Citizenship Papers?," 317.

3. Carson, "How about Citizenship Papers?," 318.

4. E. R. Kalmbach, *The Starling in the United States*, Farmers Bulletin No. 1571, U.S. Department of Agriculture, June 1931.

5. Peggy von der Goltz, "A Bird Named Jesse James," *Washington (DC) Evening Star*, March 23, 1941.

6. Teale, "In Defense."

7. C. Fred Bodsworth, "The Starling—Saint or Sinner?," *Maclean's*, January 15, 1949, 36.

8. Bodsworth, "Starling—Saint or Sinner?," 36.

9. Samuel Pickering Jr., "The Starling Is a Real American," *New York Times*, March 1, 1981.

10. Walter D. Koenig, "European Starlings and Their Effect on Native Cavity-Nesting Birds," *Conservation Biology* 17, no. 4 (August 2003): 1134–40.

11. Lyanda Lynn Haupt, "Even If We Don't Love Starlings, We Should Learn to Live with Them," TED Ideas, June 20, 2017, https://ideas.ted.com/even-if-we-dont-love-starlings-we-should-learn-to-live-with-them/.

14. How to Kill a Starling

1. E. R. Kalmbach, *Suggestions for Combatting Objectionable Roosts of Birds with Special Reference to Those of Starlings*, Wildlife Leaflet No. 172 (Washington DC: U.S. Department of the Interior, U.S. Fish and Wildlife Service, December 1940), 9.

2. Kalmbach, *Suggestions*, 9.

3. Kalmbach, *Suggestions*, 10.

4. Kalmbach, *Suggestions*, 11.

5. Kalmbach, *Suggestions*, 15–16.

6. Kalmbach, *Suggestions*, 6.

7. "Owl Devastates Starlings," *Washington (DC) Evening Star*, February 8, 1934.

8. "3,000 Starlings Killed in Lima," *Bluffton (OH) News*, August 7, 1941.

9. "Stuffed Owl in Electric Cage Finally Solves Starling Problem," *Washington (DC) Evening Star*, March 27, 1937.

10. Teale, "In Defense"; "High Pitch Radio Waves Drive Off Starlings," *Schenectady Gazette*, October 16, 1952.

11. "Vast Starling Flocks Ruin Midwest Crops," *Crowley (LA) Post Signal*, January 28, 1960.

12. Charles Fergus, *The Wingless Crow* (University Park PA: Penn State University Press, 2007), 38; "Starlings Find Few Friends in Waterbury," *New York Times*, December 14, 1973.

13. "Bygone Muncie: The Great 20th Century Starlings War," *Muncie (IN) Star Press*, July 29, 2022.

14. Fergus, *Wingless Crow*, 39.

15. Fergus, *Wingless Crow*, 39.

16. "Battle of the Starlings to Start from Scratch with Itch Powder," *Washington (DC) Evening Star*, January 25, 1948.

17. "Starlings at White House—AARWK!" *Boston Globe*, November 10, 1962.

18. "Starlings at White House."

19. "Starlings at White House."

15. Blast 'Em with Starling Calls

1. Hubert Frings, Joseph Jumber, and Mable Frings, "Studies on the Repellent Properties of the Distress Call of the European Starling (Sturnus vulgaris)," Occasional Papers from the Department of Zoology and Entomology, No. 55 (State College: Pennsylvania State University, 1955), 8.

2. Hubert Frings and Joseph Jumber, "Preliminary Studies on the Use of a Specific Sound to Repel Starlings (*Sturnus vulgaris*) from Objectionable Roosts," *Science* 119, no. 3088 (March 5, 1954): 318.

3. Frings and Jumber, "Preliminary Studies," 318.

4. Frings, Jumber, and Frings, *Studies on the Repellent Properties*, 8.

5. Frings, Jumber, and Frings, *Studies on the Repellent Properties*, 8.

6. Frings, Jumber, and Frings, *Studies on the Repellent Properties*, 9.

7. Frings, Jumber, and Frings, *Studies on the Repellent Properties*, 15.

8. Erwin Pearson, Paul R. Skon, and George W. Corner, "Dispersal of Urban Roosts with Records of Starling Distress Calls," *Journal of Wildlife Management* 31, no. 3 (July 1967): 505.

9. C. C. Miniclier, "Records Are for the Birds, but They Sell," *Santa Cruz (CA) Sentinel*, November 2, 1977.

16. Darkness in the Golden State

1. Stanley G. Jewett, "European Starling in California," *Condor* 44, no. 2 (March–April 1942): 79.

2. Charles C. Siebe, "Starlings in California," *Proceedings of the 2nd Vertebrate Pest Conference* 2, no. 2 (March 1964): 40.

3. Mark Knight, "Battle Lines Drawn for War on Starlings," *San Bernadino (CA) Sun*, September 8, 1960.

4. William C. Harrison, "The Starlings: Million Dollar Pests," *Santa Cruz (CA) Sentinel*, July 18, 1965.

5. Knight, "Battle Lines Drawn."

6. USDA–APHIS–Wildlife Services, "The Use of DRC-1339 in Wildlife Damage Management," chap. 17 in *Human Health and Ecological Risk Assessment for the Use of Wildlife Damage Management Methods* (Washington DC: U.S. Department of Agriculture, 2019).

7. Robert G. Schwab, ed., *Starling Control Research in California: Progress Report for 1966* (Davis: University of California Agriculture Experiment Station, USDA Bureau of Sport Fisheries and Wildlife, and California Department of Agriculture, 1966).

8. S. P. Garg, A. Zajanc, and R. A. Bankowski, "The Effect of Cobalt-60 on Starlings (*Sturnus vulgaris vulgaris*)," *Avian Diseases* 8, no. 4 (November 1964): 555–61.

9. W. C. Royall Jr., J. L. Guarino, A. Zajanc, and C. C. Siebe, "Movements of Starlings Banded in California," *Bird Banding* 43, no. 1 (January 1972): 26–37.

10. "Scourge from the Sky, Broadcast No. 7133," University of California Radio-Television Administration, October 3, 1965.

17. Rise of the Bird Men

1. "Two-Faced Owls Spreading to Many U.S. Communities," *Battle Creek (MI) Enquirer*, September 25, 1947.

2. "Two-Faced Owls."

3. Joseph Garretson, "Starling Strategy," *Cincinnati Enquirer*, September 23, 1947.

4. Garretson, "Starling Strategy."

5. "Owls Scare Starlings," *Life*, December 29, 1947.

6. "Soules Saves Church from Pigeon Invaders," *Decatur (IL) Herald Sun*, November 14, 1954.

7. Bob Greene, "Jimmie Soules; Suite Is What It's All About," *Chicago Tribune*, November 15, 1983.

8. "Bird Psychologist Jimmie Soules Starts New Campaign," *Decatur (IL) Herald*, January 3, 1954.

9. "Birdman's Magic Black Box Chasing St. Louis Pigeons," *Evansville (IN) Press Sun*, April 19, 1954.

10. "Birdman Soules Finds He Can't Slow Down," *Decatur (IL) Herald and Review*, March 12, 1972.

11. John Thomas, "Joseph Fink, Worst Friend of Pigeons, Gains Place in World of Celebrities," *South Bend (IN) Tribune*, June 22, 1975.

12. "London's Birds May Get Hotfoot," *Fort Lauderdale (FL) News Sun*, October 6, 1963.

13. Thomas, "Joseph Fink."

14. Mary McGregory, "Starlings on the Potomac," *New York Times*, January 20, 1957.

15. "London's Birds."

16. Thomas, "Joseph Fink."

17. "Bird in Hand Can Be a Mess," *Boca Raton (FL) News*, December 29, 1972.
18. "Against the Birds," *Sports Illustrated*, December 16, 1968.
19. Quoted in William F. Woo, "Something for the Birds," *St. Louis Post-Dispatch*, December 6, 1964.
20. "GB Birdman Standke Flying Hight," *Great Bend (KS) Tribune*, February 1, 1962.
21. G. H. Dyle, "The Great Starling Chase," *Elks Magazine*, January 1965.
22. Quoted in "GB Birdman Standke."
23. Woo, "Something for the Birds."
24. Fred D. Cavinder, "The Birds for the Birds," *Indianapolis Star*, June 17, 1973.
25. Johnson A. Neff to a colleague named Ken, January 21, 1965, archives of the National Wildlife Research Center, USDA–Wildlife Services, Fort Collins CO.
26. Neff to Ken, January 21, 1965.
27. Cavinder, "Birds for the Birds."
28. Woo, "Something for the Birds."

18. Death from Above

1. Seymour R. Linscott, "Blame Starlings for Crash," *Boston Globe*, October 6, 1960.
2. "The Fatal Starlings," *Time*, October 24, 1960.
3. Richard A. Dolbeer, "The History of Wildlife Strikes and Management at Airports," in *Wildlife in Airport Environments: Preventing Animal-Aircraft Collisions through Science-Based Management*, ed. Travis L. DeVault, Bradley F. Blackwell, and Jerrold L. Belant (Baltimore: Johns Hopkins University Press. 2013), 1–6.
4. "Aircraft Accident Report," File no. 1-0043, *Civil Aeronautics Board*, July 31, 1962.
5. "Aircraft Accident Report."
6. Michael N. Kalafatas, *Bird Strike: The Crash of the Boston Electra* (Waltham MA: Brandeis University Press, 2010), 59.
7. John J. Swearingen and Stanley R. Mohler, *Sonotropic Effects of Commercial Air Transport Sound on Birds* (Oklahoma City: Federal Aviation Agency, Aeronautical Center, 1962), 2.
8. Swearingen and Mohler, *Sonotropic Effects*, 2.
9. "A Test to Determine If Bobwhite Quail Hunt Crickets by Sound," *Auk* 86 (April 1969): 348–49.
10. Victor Ferrari Jr., "Honk! Warble! Tweet! Facts on Bird Strikes," *Aerospace Safety* 23, no. 11 (November 1967): 8–9.
11. "Giant Flashbulb for the Birds," *Santa Cruz (CA) Sentinel*, February 3, 1966.
12. Dolbeer, "History of Wildlife Strikes"; Michael J. Begier, Richard A. Dolbeer, and Jenny E. Washburn, *Protecting the Flying Public and Minimizing*

Economic Losses within the Aviation Industry (Washington DC: USDA–Wildlife Services, 2019).

13. Scott C. Barras, Sandra E. Wright, and Thomas W. Seamans, *Blackbird and Starling Strikes to Civil Aircraft in the United States, 1990–2001*, Staff Publications 200 (Washington DC: USDA National Wildlife Research Center, 2003).

14. "Seagulls Were Shot at JFK / Birds Pose Threat to Plane Engines," *Greensboro News and Record*, January 11, 1992.

15. "Thousands of Birds Delayed Morning Flights at Asheville Regional Airport," *Asheville (NC) Citizen Times*, October 25, 2019.

19. Can't Beat 'Em? Eat 'Em

1. "House Members Smack Lips during Starling Pie Feast," *Washington (DC) Evening Star*, March 16, 1934.

2. "House Members Smack Lips."

3. "Swat the Starling Slogan Is Offered Along with Recipe," *Washington (DC) Evening Star*, March 26, 1934.

4. "Recipe Is Offered for Starling Pie to End Pestilence," *Washington (DC) Evening Star*, February 13, 1934.

5. Bready, *European Starling*, 29.

6. John M. Phillips and Jack Miner, "The Starling Menace," *Pennsylvania Game News* 11, no. 10 (January 1940): 31.

7. Phillips and Miner, "Starling Menace," 7.

8. Phillips and Miner, "Starling Menace," 8.

9. Phillips and Miner, "Starling Menace," 7.

10. "Starling Stout," *Sports Illustrated*, April 23, 1956.

11. "The Rambler Considers Starling Pie," *Washington (DC) Evening Star*, November 21, 1961; *merle* is the French word for "blackbird."

12. Barry Broadfoot, "Why Not Try a Delicious Starling Pie?," *Maclean's*, November 5, 1966.

13. Earl Baysinger, "The Starling: An Untapped Resources," internal memorandum, USDA-APHIS, September 13, 1971, and subsequent memos from various USDA-APHIS employees, archives of the National Wildlife Research Center, USDA–Wildlife Services, Fort Collins CO.

14. Memos from USDA-APHIS employees.

15. "Any Way You Cook It, It's Still a Starling," *Billings (MT) Gazette*, December 2, 1964.

16. Memos from USDA-APHIS employees.

17. Memos from USDA-APHIS employees.

18. Richard Dolbeer, "Ohio's Changed Wildlife Resources: Passenger Pigeons to Starlings," keynote speech, Symposium on Regional Wine and Foods, Huron OH, December 9, 1992.

19. Joe Roman, "Eat the Invasives!," *Audubon*, September 2004.

20. Roman, "Eat the Invasives!"

21."Filleting the Lion," National Ocean Service, National Oceanic and Atmospheric Administration, 2015, https://oceanservice.noaa.gov/news/lionfish/eatlionfish.html.

22.Cassandra Profita, "'Tastes like Chicken': An Invasive Species BBQ," *Oregon Field Guide*, Oregon Public Broadcasting, February 11, 2016.

20. Mapping the Travelers

1.Olin Bray to Rodney Wiebe, July 19, 1971, archives of the National Wildlife Research Center, USDA–Wildlife Services, Fort Collins CO; Olin E. Bray, Kenneth H. Larsen, and Donald F. Mott, "Winter Movements and Activities of Radio-Equipped Starlings," *Journal of Wildlife Management* 39, no. 4 (October 1975): 795–801.

2.Ian Newton, *The Migration Ecology of Birds* (London: Academic Press, 2008), 231–32.

3.Newton, *Migration Ecology of Birds*, 231–32.

4.Harold B. Wood, "The History of Bird-Banding," *Auk* 62 (April 1945): 256–65.

5.Olin E. Bray, "Radiotelemetry for Studying Problem Birds," *Bird Control Seminars Proceedings* 122 (1973): 198.

6.Edward S. Thomas, "A Study of Starlings Banded at Columbus, Ohio," *Bird-Banding* 5, no. 3 (July 1934): 119.

7.Lawrence E. Hicks, "Individual and Sexual Variations in the European Starling," *Bird-Banding* 5, no. 3 (July 1934): 103.

8.Thomas, "Study of Starlings," 122.

9.Thomas, "Study of Starlings," 122.

10.Brina Kessel, "Distribution and Migration of the European Starling in North America," *Condor* 55, no. 2 (March–April 1953): 62.

11.Theunis Piersma, A. H. Jelle Loonstra, Mo A. Verhoeven, and Thomas Oudman, "Rethinking Classic Starling Displacement Experiments: Evidence for Innate or for Learned Migratory Directions?," *Avian Biology* 51, no. 5 (February 2020): 1.

12.H. Jeffrey Homan, James R. Thiele, Garret W. Unrein, Shannon M. Gaukler, and Anthony A. Slowik, *Band Encounters of Wintering European Starlings Captured in Kansas, Nebraska, and Texas*, Staff Publications 1149 (Washington DC: USDA National Wildlife Research Center, 2012), 124.

21. Poison Years

1.Ann Cottrell Free, "Bye Bye Blackbird: Army Sprays to Kill in 'Operation Hysteria,'" *Defenders*, June 1975.

2.Wayne King, "Half-Million Birds Killed at Army Base in Kentucky," *New York Times*, February 21, 1975.

3.King, "Half-Million Birds Killed."

4.Paul W. Lefebvre and John L. Seubert, "Surfactants as Blackbird Stressing Agents," *Proceedings of the 4th Vertebrate Pest Conference* (February 1970): 156–61.

5. Frank A. Stubblefield, "Extension of Remarks," *Proceedings of the U.S. House of Representatives* (December 10, 1974): 28992.

6. Wayne King, "Five Million Birds Darken Army Post," *New York Times*, February 9, 1975.

7. *Blackbird and Starling Control: Hearing on H.R. 11510 by the U.S. House of Representatives, Subcommittee on Fisheries and Wildlife Conservation and the Environment*, 94th Cong. 9 (1976) (statement of Wendell H. Ford, senator from Kentucky).

8. Wayne King, "Blackbirds, Roosting by the Millions, Damaging Farms in South," *New York Times*, February 29, 1976.

9. King, "Blackbirds, Roosting."

10. King, "Five Million Birds."

11. *Blackbird and Starling Control*, 12 (statement of Robin Beard, representative of Tennessee).

12. *Blackbird and Starling Control*, 86–91 (statement of Bruce Terris, attorney for the Society for Animal Rights).

13. Free, "Bye Bye Blackbird."

14. "President's Page," *South Shore Audubon Society Newsletter* 5, no. 8 (April 1975).

15. National Audubon Society, quoted in "President's Page."

16. *Blackbird and Starling Control*, 142 (statement of Floyd Ford, professor of biology, Austin Peay State University).

17. King, "Blackbirds, Roosting."

18. Richard Dolbeer, Donald F. Mott, and Jerry Belant, "Blackbirds and Starlings Killed at Winter Roosts from PA-14 Applications, 1974–1992: Implications for Regional Population Management," *Proceedings of Eastern Wildlife Damage Management Conference* 7 (1997): 77–86.

22. A Forever War

1. Victor Epstein, "Bird Culling Fallout Alarms Central NJ Community," *Morning Call* (Allentown PA), January 27, 2009.

2. USDA–APHIS–Wildlife Services, "Use of DRC-1339."

3. H. Jeffrey Homan, Ron J. Johnson, James R. Thiele, and George M. Linz, "European Starlings," *Wildlife Damage Management Technical Series* (Washington DC: U.S. Department of Agriculture, September 2017).

4. Mike Stark, "Shock and Caw: Pesky Starlings Still Overwhelm," Associated Press, September 15, 2009.

23. Spellbound

1. Jonathan Rosen, "Flight Patterns," *New York Times*, April 22, 2007.

2. Viewer comment on Sophie Windsor Clive and Liberty Smith, "Murmuration," YouTube video, December 2, 2011, https://www.youtube.com/watch?v=iRNqhi2ka9k.

3. Andy Morris, "Educational Landscapes and the Environmental Entanglement of Humans and Non-Humans through the Starling Murmuration," *Geographical Journal* 185, no. 3 (2019): 8.

4. Søren Solkær, "Gazing at the 'Black Sun': The Transfixing Beauty of Starling Murmurations," *New York Times*, April 4, 2022.

5. Solkær, "Gazing at the 'Black Sun.'"

6. Cooke, *Spread of the European Starling*.

7. Noah Strycker, *The Thing with Feathers: The Surprising Lives of Birds and What They Reveal about Being Human* (New York: Riverhead Books, 2014), 38.

8. Strycker, *Thing with Feathers*, 38–39.

9. Strycker, *Thing with Feathers*, 39.

10. Andrea Cavagna and Irene Giardina, "The Seventh Starling," *Significance* 5 (June 2008): 66.

11. Cavagna and Giardina, "Seventh Starling," 66.

12. Andrea Procaccini, Alberto Orlandi, Andrea Cavagna, Irene Giardina, Francesca Zoratto, Daniela Santucci, Flavia Chiarotti, et al., "Propagating Waves in Starling, *Sturnus vulgaris*, Flocks under Predation," *Animal Behavior* 82, no. 4 (October 2011): 759–65.

13. R. F. Storms, C. Carere, F. Zoratto, and C. K. Hemelrijk, "Complex Patterns of Collective Escape in Starling Flocks under Predation," *Behavioral Ecology and Sociobiology* 73, no. 10 (2019): 1–10.

14. Anna Azvolinsky, "Birds of a Feather . . . Track Seven Neighbors to Flock Together," Princeton University, Office of Engineering Communications, February 7, 2013, https://www.princeton.edu/news/2013/02/07/birds-feather-track-seven-neighbors-flock-together.

15. David Rosenberg, "Elegant Swarms of Starlings," *Slate*, February 21, 2014.

24. Built for Survival

1. Corey T. Callaghan, Shinichi Nakagawa, and William K. Cornwell, "Global Abundance Estimates for 9,700 Bird Species," *PNAS* 118, no. 2 (2021): 1–10.

2. Pat Leonard, "Starling Success Traced to Rapid Adaptation," *Cornell Chronicle*, February 9, 2021.

3. Jean-Nicholas Audet, Mélanie Couture, and Erich D. Jarvis, "Songbird Species That Display More-Complex Vocal Learning Are Better Problem-Solvers and Have Larger Brains," *Science* 381, no. 6663 (September 14, 2023): 1170–75.

4. Erich D. Jarvis, "Vocal Learning Linked to Problem Solving Skills and Brain Size," Rockefeller University, news release, September 14, 2023.

5. P. P. Bitton and B. A. Graham, "Change in Wing Morphology of the European Starling During and After Colonization of North America," *Journal of Zoology* 295 (2015): 254–60.

6. University of Mount Saint Vincent, "The Engine of Evolution," November 18, 2020.

7. Kenneth V. Rosenberg, Adriaan M. Dokter, Peter J. Blancher, John R. Sauer, Adam C. Smith, Paul A. Smith, Jessica C. Stanton, Arvind Pajnjabi, Laura Helft, Michael Parr and Peter P. Marra, "Decline of the North American Avifauna," *Science* 366 (October 2019): 120–24.

8. Jim Daley, "Silent Skies: Billions of North American Birds Have Vanished," *Scientific American*, September 19, 2019.

9. Elizabeth Preston, "The Day We Didn't Save the Starling," *Audubon*, November 2, 2022.

Epilogue

1. "City at War on Starlings," *New York Times*, July 30, 1914; "Starlings Stop Clock in Hartford, Conn." *Washington Times*, December 15, 1938.

2. S. Lowe, M. Browne, S. Boudjelas, and M. De Poorter, *100 of the World's Worst Invasive Alien Species: A Selection from the Global Invasive Species Database* (Auckland, New Zealand: Invasive Species Specialist Group, 2000).

3. Intergovernmental Platform on Biodiversity and Ecosystem Services (IPBES), *Thematic Assessment Report on Invasive Alien Species and Their Control* (Bonn, Germany: IPBES Secretariat, September 2023): 1–54.

4. "Starling Control: How to Get Rid of Starlings," Bird Barrier, accessed June 7, 2024, https://birdbarrier.com/starling-control.

5. Bradley F. Blackwell, Thomas W. Seamans, Morgan B. Pfeiffer and Bruce N. Buckingham, "European Starling Nest-Site Selection Given Enhanced Direct Nest Predation Risk," *Wildlife Society Bulletin* 45, no.1 (February 2021): 62–69.

6. James R. Thiele, "One Step Closer to a Better Starling Trap," *Human-Wildlife Interactions* 14, no. 3 (Winter 2020): 419–26.

7. Ghazi Mahjoub, Mark K. Hinders and John P. Swaddle, "Using a 'Sonic Net' to Deter Pest Bird Species: Excluding European Starlings from Food Sources by Disrupting Their Acoustic Communication," *Wildlife Society Bulletin* 39, no. 2 (June 2015): 326–33.

8. Rita Price, "Woman Hopes Missing Starling Answers Cellphone Call," *Columbus Dispatch*, December 19, 2019.

9. Chapman, "European Starling," 479.

10. Chapman, "European Starling," 484.

11. Chapman, "European Starling," 484.

12. Chapman, "European Starling," 485.

13. Chapman, "European Starling," 485.

14. Emily Dickinson, "Hope," in *The Complete Poems of Emily Dickinson*, ed. Thomas H. Johnson (Cambridge MA: Belknap Press of Harvard University Press, 1951), 116.

www.ingramcontent.com/pod-product-compliance
Lightning Source LLC
Chambersburg PA
CBHW021245170125
20390CB00004B/4